Environment, Health and Safety Governance and Leadership

Environment, health and safety (EHS) management has become increasingly important in the past 10 years, especially within high risk and high reliability organizations. EHS is driven from the top of an organization, and whilst there has been much research on the subject of EHS leadership, there is very little on EHS governance and the director's role in leading or influencing change in organizational safety/EHS performance.

Environment, Health and Safety Governance and Leadership: The Making of High Reliability Organizations reviews the factors influencing safety/EHS leadership and governance and addresses all the areas where the role impacts on the performance and sustainability of organizations. Based on the author's in-depth research, the book draws on much of the best-practice standards developed by many leading organizations such as the UK Health and Safety Executive (HSE), the Institute of Directors (IoD) and the Organisation for Economic Co-operation and Development (OECD).

This book provides exclusive insights and legal imperatives for practitioners and leaders to inform decision making, strategy and EHS governance, all of which can have a fundamental impact on business continuity, developing company value and the sustainability of large organizations around the world.

Waddah S. Ghanem Al Hashmi is the Executive Director, Environment, Health, Safety, Security, Quality and Corporate Affairs, for the ENOC Group in Dubai, United Arab Emirates. He is the Chairman of various committees within the organization, including the Wellness and Social Activities Program, which serves over 7,000 employees. He is also the Vice Chairman of the Dubai Centre for Carbon Excellence, PJSC, "Dubai Carbon" since 2010. He is a Fellow of the Energy Institute, an Associate Fellow of the Institution of Chemical Engineers, a member of the American Society of Safety Engineers and a Member of the Institute of Directors in the UK.

'Dr Waddah S. Ghanem Al Hashmi's book superbly brings together in one place the wealth of theory and practice that has developed around EHS governance and leadership in the past twenty years. His analysis of the key facets critical to the success of business leaders in carrying out their EHS roles and responsibilities is founded on his through and exemplary review of the evidence. This book is a must read for leaders of organisations in the high and medium hazard sectors. It provides clearly and succinctly the answers to the three key questions: "What should I be doing to ensure my organisation is highly reliable? Why? And how?" As an active proponent of the importance of active and influential top level leadership over the past twenty years, I am so pleased to note that Dr Al Hashmi's book answers these questions emphatically.'

Neal Stone, *formerly Policy Director,*
British Safety Council

'Many catastrophic safety and environmental events in high risk industries have been attributed to lack of leadership. Dr Waddah's book makes a much-needed contribution on how senior leaders can move their organisations from high risk to high reliability; safeguarding workers and shareholder value in the process.'

Teresa Budworth, *Chief Executive, NEBOSH*

Environment, Health and Safety Governance and Leadership

The Making of High Reliability Organizations

Waddah S. Ghanem Al Hashmi

Routledge
Taylor & Francis Group

LONDON AND NEW YORK

First published 2018
by Routledge
2 Park Square, Milton Park, Abingdon, Oxon OX14 4RN

and by Routledge
711 Third Avenue, New York, NY 10017

Routledge is an imprint of the Taylor & Francis Group, an informa business

British Library Cataloguing-in-Publication Data
A catalogue record for this book is available from the British Library

Library of Congress Cataloging-in-Publication Data
Names: Shihab Ghanem Al Hashmi, Waddah, author.
Title: Environment, health and safety governance and
    leadership : the making of high reliability organizations /
    Dr. Waddah S. Ghanem Al Hashmi.
Description: Abingdon, Oxon ; New York, NY : Routledge, 2018. |
    Includes bibliographical references and index.
Identifiers: LCCN 2017025921 (print) | LCCN 2017026809 (ebook) |
    ISBN 9781315713427 (eBook) | ISBN 9781138888456 (hbk : alk. paper)
Subjects: MESH: Occupational Health | Safety Management | Industry |
    Organizational Policy | Social Responsibility | Models, Organizational
Classification: LCC R859.7.S43 (ebook) | LCC R859.7.S43 (print) |
    NLM WA 400 | DDC 610.289—dc23
LC record available at https://lccn.loc.gov/2017025921

ISBN: 978-1-138-88845-6 (hbk)
ISBN: 978-1-315-71342-7 (ebk)

Typeset in Bembo
by Apex CoVantage, LLC

I dedicate this work to the fraternity of EHS practitioners as much as I do to the leaders of high risk organizations around the world. I believe this is an important and insightful work that will help bring about greater understanding and appreciation of the critical importance of EHS leadership and governance.

I could not have done this work without the perseverance and support of my beloved wife; she has been a source of inspiration and support and has continued to fuel my passion with her unconditional love and sacrifice.

Finally, to my father and mother (may she rest in eternal peace), for instilling the value of love and respect for knowledge and education. May the Lord shower them both with His endless and infinite mercy and grant them the highest of orders in Paradise for their unconditional love, righteous upbringing of me and support throughout my life.

Contents

Acknowledgements

I would like to thank many individuals for their great support throughout my research work before and during the development of this book. First I would like to acknowledge all the leaders who gave their time to me and agreed to be surveyed and interviewed in the past few years.

I have been more than privileged, even blessed, to have had two great, passionate, highly supportive and knowledgeable supervisors during my doctoral research. Professor Jackie Ford has taken a great deal of time not only to review my work but to continually and relentlessly engage me to challenge my thinking and emerging thoughts. Much of the reason I had done a DBA in the first place is attributed to Professor Nancy Harding's encouragement and continued support. She, like Professor Jackie Ford, has played a significant role in driving me to think, explore, synthesize and write.

I would also like to acknowledge the invaluable input of Professor Domenic Cooper, CEO, B-Safe Management Solutions Inc., who is one of the leading authorities worldwide in safety leadership and safety cultures. We had many insightful dialogues on safety leadership and organizational EHS culture development, methodology, methods and analysis in the past few years.

Finally, I wish to also thank the following people for their support and facilitation:

* Mr Eddie Morland, former CEO of HSL in the UK;
* Dr Bill Nixon, former Director of the UK-HSE;
* Mrs Teresa Budworth, CEO of NEBOSH in the UK;
* Mr Chris Garton, former Operations Manager, Dubai International Airport;
* Mr Peter Mohring, General Manager, Airport Navigation Services (Serco), in Dubai;
* Mr Gerard Forlin, QC – Cornerstone Barristers, London, UK;
* Dr Chitram Lutchman of Safety Erudite in Canada;
* Mr Ahmed Khalil Ebrahim, Fire and Safety Manager at BAPCO in the Kingdom of Bahrain;
* Mr Rahamat of the Barik Group in the Sultanate of Oman;
* And Mr Kush Srivastav, who helped me in reviewing this book.

Finally, I would like to place on record my gratitude to a man who believed in me and supported me in more ways than one throughout the last decade of my career, Mr Saif Al Falasi, Group CEO (since 2015) of the Emirates National Oil Company in Dubai, UAE.

About the author

Dr Waddah S. Ghanem Al Hashmi
BEng (Hons), MBA, MSc, AFIChemE, FEI, MIoD

Born in 1975, Dr Waddah S. Ghanem Al Hashmi graduated from the University of Wales College Cardiff, School of Engineering, Department of Materials and Minerals, with a Bachelor of Engineering Degree (Honors) in Environmental Engineering. He originates from a family of scholars, thinkers, statesmen and, most importantly, poets.

Waddah Ghanem is currently the Executive Director of EHSSQ & Corporate Affairs at the ENOC Group in Dubai, United Arab Emirates. In this capacity, Waddah is responsible for overseeing the development and implementation of best practices and standards for EHS, Business Excellence & Quality, Sustainability, Security and Risk Management as well as Wellness and Social Affairs and advocates the continued improvement of the company's EHSSQ culture through his leadership of the company's EHSSQ regional team. Waddah also oversees the Group's Legal Affairs as well as the development and implementation of the ENOC Group Communication Strategy to support commercial growth and strategic positioning.

Waddah initially joined ENOC as Environmental, Health, Safety and Security Supervisor in the ENOC Processing Company LLC (ENOC's Refinery) and before that worked with the international technical consulting firm Hyder Consulting. Waddah has spent the last 17 years moving from EHS Supervisor at the refinery to assistant EHS Advisor in the Group, later to grow the EHS Compliance function until he became Director, EHSQ Compliance, in 2010. In 2015 he was promoted to Executive Director, EHSSQ & Corporate Affairs. Waddah is also the Vice Chairman of the Board of Directors of Dubai Carbon in which ENOC Group is an active shareholder.

Waddah is an experienced engineering professional with in-depth expertise in environment, health and safety, risk management and sustainability issues. Trained as an environmental engineer who became a consultant early on in his career, he then diversified his career by moving into the energy industry by joining ENOC Processing Company LLC.

He is also the chairman of the ENOC Wellness and Social Activities Program Committee for the ENOC Group. In addition, Waddah chairs and is a member of various organizational committees such as the Energy and Resource Management Technical Committee and the Marine EHS Committee.

Waddah holds two Diplomas in Environmental Management and Safety Management from the United Kingdom, an MSc in Environmental Sciences from the UAE University, and an Executive MBA from University of Bradford in the UK. His recent doctoral research focused on Corporate Governance and Leadership and was also completed through the University of Bradford in the UK. He has co-authored and published several technical books and papers in the past five years including five books relating to *Safety Management* (2010), *Reflective Learning* (2014); *Operational Excellence* (2015); *Al-Fatihah: The Opening of the Quran* (2016) and the *Ten Step MBA for HSE Practitioners* (2017), respectively.

His doctoral research focused on corporate governance and leadership in high reliability organizations, particularly focusing on the GCC region.

He is a Fellow of the Energy Institute, an Associate Fellow of the Institution of Chemical Engineers and Member of the Institute of Directors (IoD) in the United Kingdom as well as a member of the American Society of Safety Engineers (ASSE).

He is based in Dubai, United Arab Emirates, is married and is blessed with five children.

Illustrations

Figures

Tables

Abbreviations

ACC:	American Chemical Council
ADNOC:	Abu Dhabi National Oil Company
ASSE:	American Society of Safety Engineers
BAPCO:	Bahrain Oil Company
BoD:	Board of Directors
BP:	An international oil major previously known as British Petroleum
BSC:	Balanced Score Card
CCPS:	Centre for Chemical Process Safety
CEO:	Chief Executive Officer
CFO:	Chief Finance Officer
CG:	Corporate Governance
CMA 2007:	Corporate Manslaughter Act 2007 (UK Law)
COSO:	Committee of Sponsoring Organizations of the Treadway Commission
CSR:	Corporate Social Responsibility
DNV:	Det Norske Veritas
E&P:	Exploration and Production
EHS:	Environment, Health and Safety
EMS:	Environmental Management Systems
ENRON:	ENRON Corporation, USA – formally Northern Natural Gas Company. Bankrupted in 2001.
ERM:	Enterprise Risk Management
ESA:	Environmental Social Accountability
FDI:	Foreign Direct Investment
GCC:	Gulf Cooperation Council
GCC-BDI:	Gulf Cooperation Council – Board Directors Institute
GDP:	Gross Domestic Product
GRI:	Global Reporting Initiative
HAZOP:	Hazard and Operability Studies
HRO:	High Reliability Organizations
HROT:	High Reliability Organization Theory
HSAWA 1975:	Health and Safety at Work Act, 1975 (UK Law)
HSC:	Health and Safety Commission

HSE:	Health and Safety Executive
HSL:	Health and Safety Laboratories
ICI:	International Chemical Industries: An international chemical manufacturing company
ILO:	International Labour Organization
IOC:	International Oil Companies
IoCA:	Institute of Chartered Accountants
IoD:	Institute of Directors (United Kingdom)
KMO:	Kaiser-Meyer-Olkin (Statistical Term)
KOC:	Kuwait Oil Company
KPI:	Key Performance Indicators
LID:	Lead Independent Director
LTI:	Loss Time Incident
MD:	Managing Director
MMR:	Mixed Methods Research
MoL:	Ministry of Labour
NAT:	Normal Accident Theory
NGO:	Non-Governmental Organizations
NOC:	National Oil Companies
O&G:	Oil and Gas
OECD:	Organisation for Economic Co-operation and Development
OH:	Occupational Health
OHS:	Occupational Health and Safety
OHSAS:	Occupational Health and Safety Audit System
PDO:	Petroleum Development Oman
PSM:	Process Safety Management
PWC:	PricewaterhouseCoopers (Consultants)
QRA:	Quantitative Risk Assessment
RC:	Responsible Care
RIDDOR:	A short-hand for a regulation in the UK on reporting injuries and fatalities
RM:	Risk Management
SA:	Social Accountability
SAI:	Social Accountability International
SEP:	Safety Excellence Program
SHEQ:	Safety, Health, Environment and Quality
SHSI:	Safety and Health Sustainability Index
SID:	Senior Independent Director
SOX:	Sarbanes-Oxley Act (United States of America Law)
SRI:	Socially Responsible Investments
TERM:	Total Error Reduction Management
UAE:	United Arab Emirates
UK:	United Kingdom
UN:	United Nations
WEF:	World Economic Forum

In the Name of Allah the Rahman, the Most Merciful "(1) Recite in the name of your Lord who created (2) Created man from a clinging substance (3) Recite, and your Lord is the most Generous (4) Who taught by the pen (5) Taught man that which he knew not."

Chapter 30; Sura 96:Verses 1–5, Noble Quran

Foreword

It was both a pleasure and an honour to be asked to write the foreword to this short, well-written and timely book.

The fourteen chapters carefully review and analyze the increasing responsibilities boards and individual directors have for general corporate governance and, in particular, for environment, health and safety.

Further, what this book also does is to set this growing trend into the unique context of the Middle East and in particular the Gulf Cooperation Council region from the perspective of detailed specific research garnered by an expert from the region who has worked there for over 20 years.

There is clearly a regulatory trend that investigators will review not only what organizational systems were in place domestically but also globally. Faults picked up in one jurisdiction must be remedied across the whole operation wherever it is situated. The courts are increasingly making that point and punishing those who do not adequately heed this message.

Industries such as oil and gas, aviation and mining need not only to learn from their own internal issues but from the lessons from the industry as a whole. Increasing scrutiny is being used to verify if boards and individual officers were undertaking their important roles with the vigour that it requires. For instance, in the UK, 46 directors and senior managers were prosecuted in 2015–16 for health and safety offences, compared to an average of 24 over the previous five years. Of these, 34 directors were found guilty or pleaded guilty. Twelve of these were imprisoned.

An investigation by one domestic regulator is often the catalyst for other countries to commence investigations in their own countries.

The possibility of the introduction of ISO45001, the new health and safety standard replacing OHSAS 18001, later in 2017 will arguably intensify this trend and make those not wishing to modernize, harmonize, reorganize and customize their systems and standards to the much higher standard more vulnerable to investigations, prosecutions and civil law claims both domestically and around the world.

This book looks at the increasing requirement for corporate governance to be at the very heart of organizations in the GCC and hopefully will be a helpful guide and aide memoire to those tasked with running the various institutions in this hugely important region.

Gerard Forlin QC
Cornerstone Barristers, 2–3 Gray's Inn Square, London
Maxwell Chambers, Singapore
Denman Chambers, Sydney

Preface

This book explains through both an in-depth review of recent literature and the perspectives of the senior leaders in high risk and high reliability organizations on EHS governance and leadership. Much of the field work that supported this research is from those operating in the Gulf Cooperation Council (GCC) region as well as globally on environment, health and safety (EHS) leadership and governance matters through interviews. Much of this book is based on around 5–6 years' worth of deep exploratory and essential research.

EHS leadership and governance has grown in importance in the past few decades. The study into the EHS leadership and governance in high risk organizations and the relationships between EHS leadership and corporate governance are greatly important due to the significance it has for organizational dynamics and performance. Its importance today is even greater due to the direct impacts on business continuity and sustainability. So this book really reviews what makes organizations high reliability organizations from the perspective of governance frameworks and leadership development.

The literature defines broadly some key areas or themes in this area of governance and leadership from an EHS context. The themes emerging include HSE knowledge and competence (of CEOs and directors); EHS/safety leadership; (enterprise) risk management; influence and accountability; developing a safety culture and communication; reporting structure and hierarchies; legal imperative for safety; operational excellence and strong integrated management systems; and monitoring of HSE performance etc. The author explains these themes and their impact throughout the book.

This book is based on the insights from many different foundational texts, practitioner reviews and industry research papers by consultants, as well as academic research. Many organizational leaders were interviewed, mainly in the GCC region, and many supported the view that the EHS performance monitoring role of the board of directors (BoD) is a critical one, but many disagreed, stating that the BoD instead should play an active role in risk management at entity level. This is an interesting and quite fundamental difference, and this book addresses the most likely reasons for these references.

Purpose and value of this book

Directors, in accordance with the IoD best practice guidelines,[1] must play a role in shaping policies, and as the primary strategic oversight supervisory body they have the key role in monitoring and directing improvements in organizational performance, which includes enterprise-wide risk management and, specifically in high risk organizations, EHS performance. They must therefore play a more proactive role in high level supervisory and investigatory roles through board committees and board meetings. This has become particularly important in the case of high risk organizations in some industries such as oil and gas, aviation, manufacturing, maritime operations and construction etc.

For the board, one of the key drivers should be to ensure that the organizations they lead are in compliance both with the statutory laws and regulations of the jurisdiction(s) in which their company operates and that the company's established policies and procedures are adequate to meet the desired objectives. There is great benefit in exploring and investigating the perceptions of the MD and CEO or the board directors in organizations regarding these roles, accountability and how EHS leadership performance could help develop a code for corporate governance of health and safety that is applicable to the local as well as international norms.

This book addresses such practices as policy-setting mechanisms; procedural controls; the relationship between health and safety and other operational/financial/commercial aspects; organizational dynamics and structure etc., as well as their place in corporate governance practice more generally.

There has been very little work undertaken to examine the perceptions of senior managers in organizations about the role that CEOs and directors play in health and safety governance and leadership. In fact there has been not enough attention in general given to EHS matters in the standard leadership and governance frameworks and leadership development programmes.

The growing importance of EHS in organizations and the significance of corporate governance standards, especially in high risk and high reliability organizations, notably in the emerging economy states, cannot be underestimated. Not only is there a scarcity of studies connecting corporate governance and EHS leadership in organizations in general, the very few standards that have been developed have been drafted mainly by specialists and then phased through consultation cycles. Very little engaged scholarship-type research has been undertaken in the areas which connect EHS, leadership and corporate governance (O'Dea and Flin [2001];[2] Olive et al. [2006];[3] Cooper [2005][4] etc.).

It is essential that we develop models that may explain the links, modalities and relationships between different drivers and perspectives that can be tested in practice. In particular it is important to understand the relationship between the first order relationships of EHS, leadership and corporate governance and the second order relationships of compliance, performance standards and expectations and the various EHS leadership models.

The author is fundamentally an EHS practitioner and has been working within the field of EHS for more than 20 years in various functional and leadership roles. As his role now extends to other connected areas in governance and leadership development, he has taken even greater interest in this domain in the past five years. Previous practitioner reviews and studies undertaken by the author led to his belief that directors need to take the time to better understand what the EHS key performance measures mean, to study the impacts as well as direct and root causes of incidents, and to become better engaged in general. It was those earlier studies that inspired the research undertaken in this book. The author hopes that this book will be a good resource for leaders to develop further their understanding of these governance principles in the management of risk in their organizations. It is equally a good resource for enterprise risk managers, EHS advisors and other such technical and enterprise risk management functional managers.

It has been established in previous research[5, 6, 7] that the drivers behind the policy-setting mechanisms are significant and may be some of the most important aspects that should be explored further with directors and CEOs. In addition, exploration of the legislative and corporate governance debates are essential to better define the right level of knowledge and engagement expected from directors in policy-setting mechanisms and how they directly impact on effective EHS performance, loss prevention and operational excellence in an organization.

In that context the author explores further these various aspects as well as elaborates on the impact of organizational structure and risk management. And whilst further research into these areas is critically important in the future, this book provides practitioners, academics and, most importantly, managers and senior leaders with very good insights. It also concludes with a model which clearly highlights both the complexity and the variability in the multidimensional areas which need to be understood in the development of a more coherent understanding of the dynamics that drive effective risk management and, more precisely, EHS leadership and governance.

References

1 Institute of Directors (IoD) (1999): *Standards for the Board – Improving the Effectiveness of Your Board (Good Practice for Directors)*. Edited by Tony Renton. IoD and Kogan Page, London.
2 O'Dea, A. and Flin, R. (2001): "Site Managers and Safety Leadership in the Offshore Oil and Gas Industry", *Safety Science* 37, 39–57, available from www.elsevier.com.

3 Olive, Claire, O'Connor, T.M. and Mannan, M.S. (2006): "Relationship of Safety Culture and Process Safety", *Journal of Hazardous Materials*, 130, available from www.elsevier.com.

4 Cooper, M. Dominic (2005): "Exploratory Analysis of Managerial Commitment and Feedback Consequences on Behavioural Safety Maintenance", *Journal of Organizational Behaviour Management*, 26, 1–41.

5 Al Hashmi, Waddah Ghanim (2012): "Assessment of Safety Culture Through Perception Studies – Using Quantitative Methods in Management Research – Case Study From the Emirates National Oil Company (Ltd) LLC Group of Companies", Working Research Paper 02/12, Presented at ADIPEC, June 2011, Abu Dhabi, UAE.

6 Al Hashmi, Waddah Ghanim (2014): "Safety Leadership and Corporate Governance – An Exploration", The 2nd HSE Forum, 11–12 Feb 2014, Muscat, Sultanate of Oman.

7 Al Hashmi, Waddah Ghanim (2014): "Senior Managers' Perceptions of the Roles and Responsibilities of the Board of Directors Towards Health and Safety in High-Risk, High Reliability Organizations in the Middle East – An Exploration Study", The ASSE-MEC PDCA 2014, 17–20 March 2014, Manama, Kingdom of Bahrain – Publication Paper in the Conference Proceedings.

1 Introduction

EHS has become more critical for organizations because of negative impacts including both direct financial effects with immediate losses and longer-term business effects such as shareholder confidence, public distrust, class-action and financial compensations and penalties.[1] However, EHS developments have also become too complex for business managers to understand at times. Even the links of incidents to their direct, related and root causes are not fully understood, as well as the cause and effect of business decisions made which may have attributed to the losses and impacts. The same probably applies to sustainability, corporate social responsibility (CSR) and enterprise risk management (ERM). Due to the growing complexity and interactivity, EHS professionals and managers/directors are finding themselves working ever more closely with the strategic business planning functions in corporations.[2]

The costs of HSE incidents to industry are great. They encompass direct costs, including medical; compensation insurance; legal fees etc. and indirect costs: uninsured costs; employee morale; time lost at work; loss of experience; economic loss due to the injured person's family; and lost time in investigations, as well as many others.[3]

To put things into global context: just as an example, between July and December 2012 alone (i.e. over a span of six months), there were six major petrochemical explosions and fires; six major and serious incidents in E&P offshore incidents; one very serious onshore incident; two serious incidents in fertilizer plants; three major and serious incidents in gas plants; and 19 refinery incidents ranging from significant to major incidents. There have been fatalities and major injuries[4] as well as combined losses amounting to many millions of dollars in direct and indirect financial and reputational impacts.

When we look at the Middle East, there have been many changes in the Arab world and the GCC states in the past 30 years. Those of the past seven years are probably most significant, partly because of the strong tides of socio-political changes with underpinning socio-economic drivers (e.g. a younger population seeking job opportunities).[5] At the same time much more focus is placed on matters such as CSR, nationalization and transparency, with national and international companies expected to clearly show their (mainly socio-economic) contribution as corporate citizens. These changes can be regarded as risks when the organizations have not prepared themselves appropriately. Other major risks are talent management matters (or human capital development), business continuity and more holistic enterprise risk management. In the energy sector alone, the expectations are that energy demand in the Gulf region will increase by 150% + by 2030, and this will place even greater pressures on energy supply companies and other associated secondary industries such as manufacturing, power and utilities, aviation and the transportation and logistics sectors.[6]

But globally, an emerging risk is the requirement to apply HSE standards; these vary between jurisdictions, and the lack of uniformity has deterred organizations from entering and operating in certain markets.[7] This has become a real problem because organizations have to balance between the application of good industry practice, which helps them manage the risks in their operations, against the local socio-economic, geo-political, regulatory and general market conditions.

Globalization has also had a significant impact, particularly in the Arab and GCC economies. Traditionally these economies have been centrally controlled, but with a breakdown of economic boundaries have come under an increased pressure due to the power of markets driven by multinationals, technology and changing economic factors. This has in turn led to the need for a more informed leadership within the major industries within the GCC, with perhaps more engaged and informed boards.[8] A major benchmarking study undertaken in the O&G sector shows that even very large national organizations such as the Abu Dhabi National Oil Company (ADNOC) have had to embrace great reforms in terms of corporate governance.[9]

In 2010 the GCC – Board Directors Institute (GCC-BDI) was established as a not-for-profit organization dedicated to making a positive impact on the economies and societies of the GCC states and region through promotion of professional directorship and raising the level of board effectiveness. The founding members are from both the financial and industrial sectors, and there are professional content partners representing four of the most well-known international business consultancies, regulatory partners and corporate affiliates representing both the financial and industrial sectors. Their workshops focus on raising board directors' awareness on matters including strategic risk management, legal imperatives for board directors and leadership matters (see www.gccbdi.org).[10]

Although the fiduciary and legal duties of directors, especially in listed companies, have become more pronounced, the reference of the same in global standards and codes of practices has been covered in detail later in the book.

This also has led to an emerging risk where organizations have to apply HSE standards, which vary in different jurisdictions, so even applying a particular "best practice" may not be suitable in a certain jurisdiction. This has even deterred organizations from deciding to enter and operate in certain markets.[7]

Carey and Patsalos-Fox (2006)[11] explain that after many serious corporate governance standards have come into effect, such as the US-based Sarbanes-Oxley Act (SOX), the demand for academics, non-profit organization executives and retired executives to be engaged as board directors has increased dramatically.

In 2012, the Organisation for Economic Co-operation and Development (OECD) developed a voluntary standard on "Corporate Governance for Process Safety – Guidance for Senior Leaders in High Hazard Industries" (PSM) that focuses on high risk industries.[12] The document concludes with a model in which leadership is the heart of the model.

It is interesting to note that progressive companies now seek to fill at least a third of the board seats with seasoned professionals and specialists with expertise in corporate social responsibility (CSR), EHS, sustainability etc.[10]

In the following chapters a review of many of the aspects discussed in this chapter is undertaken in further depth, and the implications in certain areas are elaborated on.

References

1 Lukic, Dane, Margaryan, Anoush and Littlejohn, Allison (2010): "How Organizations Learn From Safety Incidents: A Multifaceted Problem", *The Journal of Workplace Learning*, 22(7), 428–450.

2 MacLean, Richard (February 2001): "Coming to Grips With Business Reality", *Environmental Protection Magazine*, 42–44.

3 Al-Ahmari, Abdullah (February 2012): "The Relationship of Good Safety With a Good Business", ASSE-MEC-2012-21, American Society of Safety Engineers – Middle East Chapter Conference and Exhibition, Bahrain, Page 152.

4 Marsh (2013): "The 100 Largest Losses 1972–2013 – Large Property Damage Losses in the Hydrocarbon Industry", Energy Practice (Global Report), 22nd Edition, Marsh and McLennan Companies.

5 PRB Assad, Ragui and Rouddi-Fahmi, Farzaneh (2007 April): "Youth in the Middle East and North Africa (MENA): Demographic Opportunity or Challenge", Population Reference Bureau (PRB), Washington, DC.

6 McKellar, Kenneth (August 2011): "Risky Business", Interview, Published by *Oil and Gas Middle East* (Magazine).

7 Richardson, Graham (2013): "Risk in Emerging Markets", *Governance*, March(225), 6–7.

8 Major, John (2005): "Multinational Corporations and State Sovereignty: Redefining Government Responsibilities in a Global Economy", *The Gulf Challenges of the Future*, Published by The Emirates Centre for Strategic Studies and Research, UAE.

9 Booz & Co (2010): "Oil and Gas Sector and Corporate Governance Benchmarking", Booz and Company Consultant Research Report Presentation, Version 2, Beirut.

10 Gulf Cooperation Council – Board Directors Institute website, www.gccbdi.org, accessed 20th September 2016.

11 Brodeur, Andre, Buehler, Kevin and Patsalos-Fox, Michael (2010): McKinsey Working Papers 18: "A Board Perspective on Enterprise Risk Management", McKinsey.

12 Organization for the Economic Cooperation and Development (OECD) (March 2012): "Corporate Governance for Process Safety – Draft Guidance for Senior Leaders in High Hazard Industries", Draft – 1st March 2012, Environment Directorate, Joint Meeting of the Chemicals Committee and the Working Party on Chemicals, Pesticides and Biotechnology, Ref: ENV/JM/ACC(2012)1.

2 Developing high reliability organizations (HRO)

The concept of high reliability organizations rests very much on the theory that accidents can be prevented through good organizational design and management/leadership.[1] One of the most comprehensive studies undertaken in more recent years defining high reliability organizations is the HSE Laboratory (2011) study "High Reliability Organizations – A Review of the Literature". In this work the definitions start from the context of the two (competing) prominent schools of thought that seek to explain accidents in complex, high hazard organizations: **Normal Accident Theory (NAT)** and **High Reliability Organization Theory (HROT)**.[2]

In NAT, the definition is very straightforward and depicts the tight coupling of various aspects of systems and system components (e.g. people, equipment, procedures) and shows that, due to the complex relationships and interdependencies of these tightly coupled and often highly automated systems, the timing of tasks does not even allow for human intervention. When a failure occurs in one part of the system, it quickly spreads to another part of the system, and you have a massive failure. Interestingly, researchers classified petroleum/petrochemical plants such as refineries as lower risk when compared to military systems and aircrafts etc.

This theory was apparently highly criticized mainly for its failure to be consistent in accurately capturing and differentiating between the design features of systems in these industries and ignoring the conditions in which complex systems operate and do not fail. Also, others have identified the weakness in the definition of the theory itself because it is too complex to serve the purpose of simply identifying how the theory describes accident causation, etc. It can therefore be argued that it thus becomes of little value to practitioners, as it fails to advise its users and suggest how accidents can be reduced.[2]

In essence, the NAT describes the consequences of human–machine interface actions where the source of failure is more physical and/or technological. In contrast, the HROT can be described as more "in response to" uncertainty, complexity and risks and with a focus more on behavioural and socio-physical aspects.[3] The socio-physical dimension is created from the tight relationship between the human being and the machine/physical processes.

The definition of HROT addresses the criticisms of NAT. This theory focuses mainly on the position that accidents in complex systems are neither unavoidable nor invertible. This is because of the processes in place that enable high hazard organizations to effectively prevent incidents and contain catastrophic errors from actually occurring and thus maintaining a consistent record of safe operations. In fact, HRO researchers maintain a positive view regarding the nature of accidents in complex systems by arguing that organizations can become more reliable by creating or engineering a positive safety culture and reinforcing safety-related behaviours and attitudes.

What is very interesting here is that HRO researchers maintain that such **organizations are not error-free as much as they are preoccupied with failure and prevention of that failure** and how to deal with failing systems. As such, and most significantly, such organizations exhibit strong learning orientation, prioritization of safety over other goals, continual training and development and an emphasis on checks and maintaining the safety performance. To this end, they also explain that HRO perspectives have much in common with resilience engineering, which includes systems employed extensively in the aviation, petrochemical and nuclear industries. However, the HROT has also had its fair share of criticisms because it ignores the broader social and environmental contexts to learn from errors. Examples quoted include the (corporate) political implications of errors that may impact on the extent to which errors can be openly reported. It is important to understand the actual characteristics of HRO, which are summarized in Table 2.1.

One recently published definition of an HRO is one "that produces product relatively error-free over a long period of time" (OECD 2012, page 12).[4] Two key attributes are described, including having "a chronic sense of unease" and therefore lacking a sense of complacency. They as such can be described

Table 2.1 Attributes of an HRO (adapted from HSE Laboratory 2011)[2]

No	Characteristic	Implication
1	*Dynamic Leadership Shift*	Whilst decision making is hierarchical during routine periods, with clear responsibilities during emergencies, the organization migrates to a structure which leverages on the members within the organization who have the expertise.
2	*Systematic Intervention*	They manage by exception, and thus managers focus on strategic and tactical decisions and seldom interfere with operational issues, which are delegated and covered by clear processes.
3	*Learning Organization*	Climate of continuous training and learning.
4	*Multi-Communication*	Several channels are used to communicate safety critical information – timely communication of information during normal and emergency situations.
5	*Redundancy*	In-built redundancy and the provision of back-up systems in case of a failure.

as believing that an incident can happen at any time even if no incidents have taken place for a very long time. The second important attribute is to "Make strong responses to weak signals" and therefore to set a low threshold for intervention. This generally means that they will go to the extent of shutting down operations to investigate effectively more often, which, while it may mean financial losses, is seen as an essential risk control measure to prevent a much bigger potential loss.

Resilience in HRO can be engineered by incorporating the following characteristics, which include:

1   A "just culture" promoting transparency in reporting of incidents and improvements, with a great balance in favour of supporting the reporting culture and against tolerating unacceptable behaviours;
2   Management commitment, which balances the pressures of production with safety and management behaviour/allocation of resources;
3   Increased flexibility through supportive systems and empowerment;
4   A learning culture in which information is shared, regular effective training is undertaken and there is continuous development of critical safety information;
5   Preparedness through proactive safety management systems;
6   Opacity/awareness through organizational collection and analysis of information that enables the organization to identify hazards and risk early and deal with prevention; and finally
7   Resources in the form of competent staff, systems, technology and additional resources to help prevent incidents and deal with them when they happen.

Clearly there are many factors here which can have an impact, and they vary depending on the organization.[2]

This is consistent with Al Hashmi's (2012)[5] Model of Safety Culture in which awareness born from information sharing and training as well as autonomy/management support factors are all indicators of a safety culture. More is discussed with respect to safety culture later in the book (see Chapter 5) because there are similarities between safety culture and high reliability within an organization. They share many similar attributes, except that the HRO definition is probably more encompassing.

In a case study paper about Al Jubail Petrochemical Company (KEMYA) it is established that management commitment and leadership are the key drivers for the KEMYA BBS behavioural-based safety programme. It linked the safety performance and management performance to process safety management (PSM) and the Operational Integrity Management System leading to their Safety Excellence Program (SEP).[6] Operational excellence is a function of high reliability and will be discussed later.

Due to the fact that the processing, manufacturing and aviation industries etc. are not only inherently risky but also highly dynamic, and with risks arising all the time with changing conditions and environments, maintaining product quality and high reliability becomes a fundamental cornerstone of the very viability and sustainability of that business. HRO generally develops people's strengths through the actions of individuals who are highly aware and practice safe attitudes within their organizations, which in turn, over time, creates an organizational culture which can be described as a high reliability culture.

There are four fundamental organizational characteristics which help HROs control the number of incidents that take place. These include:

1   An organizational prioritization of safety and the sharing of the performance goals throughout the organization;
2   An organizational culture of reliability (as described earlier);
3   A learning organization which uses higher orders of learning to continually improve; and
4   "A strategy of redundancy beyond technology".[3]

However, there also is a real difficulty with defining HROs because of the very fact that incidents occur all the time in organizations as near-hits or, as commonly referred to in the industry, as near-misses. The majority do not eventually become an accident, i.e. these incidents do not result in loss of any kind. The question of defining HROs quantitatively raises this issue with many of the definitions, as an industry benchmark is required, be it in the aerospace, aviation, oil and gas, manufacturing or other industries. So, statistically, if a nearly accident-free performance is achieved, then perhaps that particular organization can be described as an HRO.

This also means that statistically, reliability can be "calculated," but this would be a function of uncertainty. This was the fundamental issue with the whole NAT/HRO comparison study. If and when HROs are defined statistically, there is a degree of engineering/technical/mathematical accuracy, but the factors related to organizational and social matters bring about greater uncertainty to some extent.[7] This probably explains why leadership and (organizational) culture are the heart of the OECD PSM/CG model.

Professor Andrew Hopkins (2002)[8] reviewed extensively the five characteristics of HROs defined by Weick and Sutcliffe (2001):

1   Preoccupation with failure;
2   Reluctance to simplify interpretations;
3   Commitment to resilience;
4   Sensitivity of operations; and
5   Deference to experience, with the encouragement of a fluid decision-making system which they described as producing a collective state of mindfulness, which was key.

Whilst Hopkins supported their views, he also said that the challenge in defining HROs thus lies in the very fact that a detailed enquiry looking at these five areas would be required, and this was highly dependent on the industry within a context of time.

Managers in HROs work closely with their subordinates regarding their work actions rather than just focusing on figures related to bottom-line performance. Therefore, in a way, creating learning organizations enhances performance. It is worth noting that considering the focus on performance, some researchers have found that the HRO researchers have oversimplified to some extent the complexity and difficulties that engineers and scientists face, and they have suggested an alternative systems approach to safety which tries to overcome the limitations of both the NAT and HROT theories.[7]

The human element remains critical, and weak management systems compounded with human errors is what causes incidents. Therefore, human factors must be understood very clearly in the prevention of incidents. OH performance standards, training and competency, task design; workforce rationalization and time motion studies to determine the safe manning levels; and task-human-system interaction etc. all help strengthen PSM systems.[9]

In a relatively recent and dynamic study of an oil refinery in the UK, which apparently was actively working towards achieving higher levels of reliability and safety (which was in part a reaction to the review of the Texas City explosion in 2005), the researchers identified four fundamental areas or themes emerging from both one-to-one and focus group interviews, which included training and technical competence; hazard identification and awareness; learning orientation; and strong management commitment to safety.[10] The same study concluded that whilst there were some highly progressive practices, management commitment and more importantly high levels of management visibility remained the two most critical challenges faced by organizations striving to implement high reliability practices.

Finally, if it is the case that HROs have migratory decision-making processes that allow those who are closer to an incident to react to prevent escalation, or otherwise "the empowerment within the hierarchy" of those who would be better informed (or more knowledgeable), the **leadership model within these organizations would have to be highly empowering**. However, under normal conditions it would also need to be highly engaged and have an appreciation of the risks and challenges faced within the day-to-day operations of the organization.[2]

The leadership from the top must then be very trusting, but would, within the context of operational integrity, need to expect inherently highly reliable operations. This is particularly the case with operations such as oil and gas, power and utilities, and aviation etc. This would mean that operations are managed within a reliable integrated management system where the changes in response can be rapid enough to deal with any fast-occurring developments.

This also requires that managers at all levels, starting from the top of the organization, are trained to manage and respond, and this can only be achieved through structured systems, training and drills.

In conclusion, it could be argued that a culture of high reliability must be driven by the executive management team and the board of directors, who would set the tone as an expectation. The levels of empowerment would need to be commensurate with the degree of competency, personal leadership of managers to the level of stewardship as the risk owners, and true organizational maturity that seeks operational excellence and reliability.

References

1 Bibbings, Roger (September 2010): "High Reliability", *Parting Shots*, Royal Society of Prevention of Accidents (RoSPA), Pages 55–56.
2 Health and Safety Laboratory (HSL) (2011): "High Reliability Organizations – A Review of the Literature", Health and Safety Executive Research Report – RR899, HSE Books, Sudbury, Suffolk, UK.
3 High Reliability Organizing website, www.high-reliability.org, accessed 20th September 2016.
4 Organization for the Economic Cooperation and Development (OECD) (March 2012): "Corporate Governance for Process Safety – Draft Guidance for Senior Leaders in High Hazard Industries", Draft, Environment Directorate, Joint Meeting of the Chemicals Committee and the Working Party on Chemicals, Pesticides and Biotechnology, Ref: ENV/JM/ACC(2012)1.
5 Al Hashmi, Waddah Ghanim (2012): "Assessment of Safety Culture Through Perception Studies – Using Quantitative Methods in Management Research – Case Study From the Emirates National Oil Company (Ltd) LLC Group of Companies", Working Research Paper 02/12, Presented at ADIPEC – June 2011, Abu Dhabi, UAE.
6 Al Hajri, Mohammed Q. (February 2008): "BBS Leading to Safety Excellence", ASSE-MEC-0208-38, American Society of Safety Engineers – Middle East Chapter Conference and Exhibition, Bahrain, Page 265.
7 Marais, Karen, Saleh, Joseph H. and Leveson, Nancy G. (2006): "Archetypes for Organizational Safety", *Safety Science*, 565–582.
8 Hopkins, Andrew (December 2002): "Safety Culture, Mindfulness and Safe Behaviour: Converging Ideas", National Research Centre for OHS Regulation, The Australian National University.
9 Bridges, William (February 2010): "Human Factors Elements Missing from PSM)", ASSE-MEC-2010-48, American Society of Safety Engineers – Middle East Chapter Conference and Exhibition, Bahrain, Page 392.
10 Lekka, Chysanthi and Sugden, Caroline (2011): "The Successes and Challenges of Implementing High Reliability Principles: A Case Study of a UK Oil Refinery", *Process Safety and Environmental Protection*, 89, 443–451.

3 Corporate governance and historical review of developments

The increasing importance of corporate governance was noted earlier, especially following the collapse of large organizations (such as ENRON) and investors' concerns about their investment. The systems of corporate governance and control have also come under great scrutiny in recent years, with organizational investors demanding effective controls be put in place to ensure the discipline required to prevent the risk of loss of their investments.[1] After the collapse of ENRON, Breeden (2003)[2] presented an extensive report to the US government with 78 individual recommendations covering many issues including directors' qualifications and risk management.

Recent developments have occurred which Michael Jensen (1993) explains have "changed the economic landscape as rapidly as within the 19th Century Industrial Revolution".[3] These are the rapid changes in technology and organizational aspects which completely altered many models in production and labour markets. Companies have grown in size, necessitating new forms of management[4] and more sophisticated systems; this was attributed to the growth in the size of companies and much higher production scales, which led to shareholders ceasing to manage these organizations and hiring professional managers instead. Technological advances also brought about economies of scale which contributed to this growth in organizations. In time, these professional managers moved to eventually becoming board members, and changes in a board member's role towards being an important "advisor" to the shareholder.

Another important observation is that organizations have moved over the years from being small and medium-size enterprises, and as organizations become bigger they start to require a greater degree of planning, accounting, operational management and systems in all their various functionalities.[5] This has led to the development of more regimented systems, with process mapping, procedures and checklists.

These systems are usually employed in response to a senior person within the organization and generally when a company board demands that a hierarchical and systematized organization is needed, then it is generally acceptable to assume that it is one in which better corporate monitoring and control can be exercised. This, however, may carry the threat of inhibiting innovation because many of the operations become so systematic and follow such strict guidelines and procedures that bureaucracy is brought into being.

These systems generally emerge to compensate for incompetence, inconsistencies (which cannot be tolerated in certain high risk industries such as aviation and oil and gas) and generally what can be described as a lack of discipline. To become a "great" organization, a balance between a high *ethic of entrepreneurship* and a *high culture of discipline* has to be created. It may be considered significant that "The good-to-great companies built a consistent system with clear constraints, but also gave people freedom and responsibility within the framework of that system. They hired self-disciplined people who didn't need to be managed, and then managed the system, not the people" (Collins, 2001, page 125).[6]

Governance and control systems by their very nature are constraining, or at the very least establish certain requirements and expectations that have to be fulfiled by people, who are then measured against these performance criteria. This creates accountability and even a culture of responsibility. A greater degree of clarity is demanded of managers and directors. The internal control framework provided by the Committee of Sponsoring Organizations of the Treadway Commission (COSO 2004)[8] has led to organizations in the US now leveraging this framework and recommendations **beyond financial reporting**.[7]

They must:

> (1) Develop a functional model identifying business processes (i.e. a Process Classification Framework) so that a well-defined policy and procedure framework exists, then (2) Support the framework with established processes to continually evaluate, update and communicate policy changes throughout the organization and (3) Leverage this framework consistently across the organization in support of various business processes.
>
> (Zukis et al. 2010, page 2)[7]

The "focus on fulfilling objectives through better risk management" is at the heart of the risk management process.[8] The Turnbull guidance on the Combined Code Corporate Governance requires that companies **have robust systems of internal control going beyond financial risks but looking more holistically at risks relating to the environment, health and safety and also business reputation**.[9]

Many boards have established ethics subcommittees (sometimes called supervisory committees) to oversee the behaviour of board members and ensure the highest standards of compliance.[10]

Defining corporate governance

Corporate governance has been defined differently in different countries. One definition is that corporate governance: "is to be seen to be acting responsibly, and informing all interested parties, or stakeholders, of decisions which will affect them" (Kendall and Kendall 1998, pages 18–19].[4] An alternative definition: "Governance is the process whereby people in power make decisions that create, destroy or maintain social systems, structures and processes" (McGregor 2000, page 11).[11]

Whilst others have also not given exact definitions, some have explained that it is about ensuring that "the proper standards are installed within an organization". In a survey, work respondents (managers) came with many different definitions including "Having an appropriate pay policy for senior people in industry"; "providing checks and balances to avoid excesses of top bosses;" "A set of procedures to protect the organization from fraud or loss due to poor practice"; "Providing checks on the management thus protecting shareholders"; "Curbing the worst excesses of a greedy managing class"; and "Providing a control climate suitable to the organization".[12]

One interesting quote reads:

> Instead of episodic, confrontational challenges for control, CEOs and directors will find themselves subjected to continuous, on-going scrutiny from both active investors and major long term institutional investors, who will seek to engage in substantive debate about specific corporate policies and overall corporate performance. . . . The new governance process is based on continuing dialogue and debate among key, long-term institutional and other investors about specific, substantive aspects of corporate policy.
>
> (Bain and Band 1996, pages 2–3)[12]

Another definition is "Corporate Governance is the process of serious decision-making at the controlling heart of the organization. For most practical purposes, this means the board and the CEO are the ultimate arbiters" (Leavy and McKieranan 2009, page 46).[5]

In the UK, in 1992 the government-commissioned Cadbury Committee Report on Financial Corporate Governance of Companies focused on the role of the chairman, CEO and directors (especially the role of independent directors), but it also reflects some of the aspects that relate to the non-financial aspects of the decisions made by organizations.[13]

The simplest and most comprehensive definition is: "Corporate governance is the system by which companies are directed and controlled". Boards of directors are responsible for the governance of their companies; shareholders appoint the directors and auditors and must satisfy themselves that an appropriate governance structure is in place. The responsibilities of the board include setting the company's strategic aims, providing the leadership to put them into effect, supervising the management of the business and reporting to shareholders on their stewardship. The board's actions are subject to laws and regulations (Cadbury Committee Report 1992, page 14).[14]

The Sarbanes-Oxley Act in 2002 (SOX Act) and other reforms have tried to encourage greater board involvement in the understanding of management performance and other best practices including:

1 Having independent directors in the majority;
2 Tightening the standards for independent directors;
3 Restricting the audit committee composition and expanding its responsibilities;
4 Requiring that compensation and nominating/governance committees are comprised entirely of independent directors and granted specific responsibilities;
5 Convening regular executive sessions restricted to the non-management directors;
6 Performing regular board and committee evaluations;
7 Others (Millstein and MacAvoy 2003).[15]

On initial review of the various codes that have been developed in various countries – e.g. in the UK, The Cadbury Code, 1992; then later the Turnbull Committee Report, 1999; in the USA, the Sarbanes-Oxley Act, 2002; and in South Africa, the King, I, II and III (King 2009)[16] – it is clear that the concept of corporate governance has grown in importance in the past 20 years, with greater transparency being demanded by stakeholders as well as shareholders. Countries such as South Africa, which applied an effective code of practice (see King I, King II and King III), have seen companies enjoying greater foreign direct investment (FDI), with the benefits of greater confidence of the investor. There has been much discussion and also debate in the USA between various institutions regarding the Sarbanes-Oxley Act (SOX), which was passed soon after the ENRON collapse/scandal. The act operates on a "Comply or Else" principle which has been very much argued to be ineffective in adding a holistic value to corporations. It is said that, for instance, the cost of compliance to SOX in the USA amounts to what is greater than the total write-off amounts on ENRON, World Com and Tyco combined.[16]

Corporate governance can be described through its most important facet – *organizational design and architecture*. It has three key elements: (1) The assignment of decision-making authority, i.e. who gets to make what decisions; (2) performance evaluation, i.e. how the performance of employees and their business units are measured; and (3) compensation structure, i.e. how employees (including senior managers) are rewarded or penalized.[3] However, there is also a growing appreciation of the softer issues of governance, i.e. those requiring a greater understanding of human nature and behaviour.[17] This subject is discussed later.

In studies of the variances in corporate values, structure and governance have an impact on corporate values, and there is little relationship between values and profitability.[18] However, Ayuso et al. 2007[19] drew different conclusions, finding evidence of a **positive relationship between CSR, country location,**

board diversity and stakeholder engagement and a firm's financial performance.

One possible reform proposal to boards to strengthen their purpose and make them more effective is an action table (Table 3.1) developed based on nine suggested initiatives.

Table 3.1 Initiatives Suggested to Improve Board Performance

Initiative	Description*	Implementation
1	Separation of the role of the chairman and the CEO and designating an independent director as chairman.	Shareholder/stakeholders
2	Determination of the satisfaction that management has appropriate processes in place to prepare the certification required by the SOX Act.	The certification is a legal requirement by the act – but it is good practice – action for the board of directors
3	Boards should take responsibility for their company strategy, risk management and financial reporting based on the company's business environment, challenges and opportunities and should carefully construct compensation arrangements to reward extraordinary company (not market) performance.	Action for the board of directors under clear leadership of the chairman.
4	Boards should assure themselves of the integrity of management.	Primary action by board committee's appointed chairmen with the board chairman.
5	Board should structure meetings so as to ensure that issues central to the performance of the company are given sufficient time and that management presentations concerning such issues present options and not simply reports.	Board secretary under clear leadership of the chairman.
6	Boards should assure themselves that the board agendas prioritize and carry out the agreed upon rules of practice.	Action for the board of directors under clear leadership of the chairman.
7	A company internal auditor should be hired and report to the board directly.	Action for the board of directors under clear leadership of the chairman.
8	Boards should feel free, without the consent of management, to retain such consultants and advisors as they deem necessary to carry out their responsibilities.	Action for the board of directors under clear leadership of the chairman.
9	Boards should expand their definitions of management and establish procedures for familiarizing themselves with business leaders below the level of senior management.	Action for the board of directors under clear leadership of the chairman.

*Note as adapted from Millstein and MacAvoy (2003)[15]

Ultimately the board of directors (BoD) play a critical role in setting the operating rules for an organization because they have the ultimate responsibility for appointing the CEO and signing off on the organization's business strategy. Boards are becoming more and more concerned with sustainability issues such as environment, health and safety and social accountability, where the risk for legal non-compliance and/or failing to meet their social obligations can have a long-term lasting negative impact, quickly eroding the organization's value. As the primary shareholder representatives, they are getting to better understand that the stakeholders' position is becoming stronger, and they are under greater scrutiny in today's business world as opposed to even a decade ago.

However, some research alludes to the disappointing reality that the surveys conducted of UK directors on the HSE/IoD 2008 Code over two years showed little improvement in awareness, leadership and implementation of the code.

In fact BoD members often lack sufficient diversity to deal effectively with the shift from the exclusive focus on the shareholders' interests to a focus on meeting the expectations of a wider, more diverse group of stakeholders. As corporate sustainability has wide-ranging implications for corporate governance, the diversity of knowledge and experience of the BoD has to move from the traditional appreciation of only financial and industry-related knowledge to a wider and more stakeholder-informed competence.[20] Williams (2008)[21] explains that even investors take greater interest today in ensuring that the organizations they invest in deal positively and effectively with ethics, environmental and social issues. Socially responsible investments (SRI) are being demanded by some pension fund members, putting pressure on fund trustees to ensure that fund managers engage in SRIs.

In the next chapter, the author seeks to explain how and why leadership within the context of corporate governance is so critical and then later explores its impact on safety and EHS performance in organizations.

Even investors take greater interest today in ensuring that the organizations they invest in have a positive ethical, environmental and social impact on society. Many organizations are evaluated on their sustainability index along with other investor indices and ratios.

References

1 Dunlop, Alex (1998): *Corporate Governance and Control.* Business Skills Series, Chartered Institute of Management Accounts (CIMA) Publishing, London.
2 Breeden, Richard (August 2003): "Restoring Trust", A Report to the Hon Jed S. Rakoff, The US District Court, For the Southern District of New York on Corporate Governance For the Future of MCI, Inc.
3 Chew, Donald H. and Gillan, Stuart L. (2005): *Corporate Governance at the Cross-roads – A Book of Readings.* McGraw-Hill/IRWIN services in Finance, Insurance and Real Estate, London.
4 Kendall, Nigel and Kendell, Arthur (1998): *Real-World Corporate Governance – A Programme for Profit-Enhancing Stewardship.* PITMAN Publishing, London.
5 Leavey, Brian and Mckiernan, Peter (2009): *Strategic Leadership – Governance and Renewal.* Palgrave Macmillan Publications, United Kingdom.

6   Collins, Jim (2001): *Good to Great*. HarperCollins Publishers, Random House Tower, New York.

7   Zukis, Bob, Quan, John, Bala, Sunil and Minakawa, Xavier (2010): *Policies and Procedures for Operational Effectiveness – Enabling a Framework For Continuous Improvement. Best Practices for Making Policies and Procedures a Platform for Operational Effectiveness*. Pricewaterhouse-Coopers Publications.

8   Committee of Sponsoring Organizations of the Treadway Commission (COSO) (September 2004): "Enterprise Risk Management – Integrated Framework", Executive Summary.

9   Institute of Chartered Accountants (IoCA) (September 1999): "Internal Control – Guidance for Directors on the Combined Code", Institute of Chartered Accountants Publications (Turnbull Report).

10 Health & Safety Executive (HSE) & Institute of Directors (IoD) (2008): "Leading Health & Safety at Work – Leadership Actions for Directors and Board Members", INDG417, HSE Books. Health & Safety Executive (HSE) (2006): "Defining Best Practice in Corporate Occupational Health and Safety Governance", A Report Prepared by Acona Ltd for the Health and Safety Executive Research Report 506.

11 PricewaterhouseCoopers (PWC) (2005): "Corporate Governance", Governance, Risk and Compliance Series – Connected Thinking, Global Best Practices, PricewaterhouseCoopers Publications.

12 Mcgregor, Lynn (2000): *The Human Face of Corporate Governance*. Palgrave Publishers, London.

13 Bain, Neville and Band, David (1996): *Winning Ways Through Corporate Governance*. Macmillan Business, London.

14 Clutterbuck, David and Waine, Peter (1993): *The Independent Board Director – Selecting and Using the Best Non-Executive Directors to Benefit Your Business*. McGraw-Hill Book Company, New York.

15 The Cadbury Code (1992): *Report of the Committee on Financial Aspects of Corporate Governance, Cadbury Report*. Gee (a division of Professional Publishing Ltd), London.

16 MacAvoy, Paul W. and Millstein, Ira M. (2003): *The Recurrent Crisis in Corporate Governance*. Palgrave Macmillan, New York.

17 King, Melvyn E. (September 2009): "King (III) – (King Committee Chairman): 'King Code of Governance for South Africa 2009 – King III'", Institute of Directors, South Africa.

18 Thomsen, Steen (2005): "Corporate Governance as a Determinant of Corporate Values", *Corporate Governance*, 5(4), 1–27.

19 Ayuso, Silvia, Rodriguez, Miguel Angel, Garcia, Roberto and Arino, Miguel Angel (January 2007): "Maximizing Stakeholder's Interests – An Empirical Analysis of the Stakeholder Approach to Corporate Governance", IESE Business School – University of Navarra, Barcelona, Working Paper no. 670.

20 Dunphy, Dexter, Griffiths, Andrew and Benn, Suzanne (2003): *Organizational Change for Corporate Sustainability*. Routledge, Taylor and Francis, London.

21 Williams, Kinda Stallworth (April 2008): "The Mission Statement – A Corporate Reporting Tool With a Past, Present and Future", *Journal of Business Communication*, 45(2), 94–119. The Association for Business Communication.

4   Legal imperatives that drive EHS governance

As in other areas of any business, legal or regulatory compliance may drive organizational behaviour. The risk of not complying can have devastating impacts on the business as a whole. Therefore one of the key roles of the CEO and the BoD of an organization is, whilst remaining focused on commercial needs and growth, to act responsibly towards all stakeholders.[1] Organizations are expected to demonstrate this compliance and commitment. In fact, if and when an omission does occur, an organization and its leadership must actually demonstrate that before this omission, which may have become manifested in an incident, they have done everything reasonably possible to prevent such an occurrence, and that, more importantly, the omission was not due to a failure to comply with the statutory duties.

Health and safety laws such as the Health and Safety at Work Act, 1975 (HASWA-1975) in the UK; the UAE Labour Law (1980) and Ministerial Order 32 (1992) – UAE; and the Singaporean Health and Safety Act (1997) are just some of the examples of laws that require a degree of protection to employees' health and safety. There is thus a clear legal binding expectation of organizations to protect employees (in fact contractors, sub-contractors and the public at large are expected to be protected as well) from adverse health and safety impacts. In the UK for example that legal protection to all (including employer, employee, contractors and the public etc.) is provided through the HSAWA-1975. However, regardless of there being legislation with respect to health and safety, under international common law principles such as the duty of care, reasonable care and protection of all are expected from the employer. Even in emerging or less matured areas of the world employees are protected by the ILO conventions and common law principles.

When we address the issue of organizational behaviour and the law, we start to understand the immense complexity that is created. Common law was created to govern the actions of individuals. As such, when organizations act, one can presume that this is the action of an individual or otherwise collectively a group of individuals. In a long and very insightful discussion Metzer (1987) explains the major problem with the application of punishment to corporations.[2] He explains that under common and civil law the application of punitive damages for an organization's wrongdoing creates a dilemma.

In most cases, for a profit-making organization these costs are either eventually passed on to the consumer or otherwise borne by the shareholders, who may have not had any control over the reasons for the omission in the first place. More critically, in larger and very serious incidents where the damages can lead to corporate bankruptcy, the harm is to an even wider circle of innocents such as employees, creditors etc., and even the communities that rely on that business.

Even with criminal corporate liability the problems are no less. Once again, the punishment is mostly financial, and depending on the market conditions either the customer or the shareholders will end up paying. The underlying assumption is that organizations tend towards value maximization and therefore try to spend a great deal of effort on limiting the financial impact upon them. However, the law is moving towards punishing senior managers responsible for the wrongdoing. Forlin (2011) explains that in the first case in the UK of a successful prosecution under the Corporate Manslaughter Act of 2007 (CMA-2007), where a young unsupervised worker died as a result of working in a pit which collapsed, where the walls were not sufficiently supported for its depth, the managing director was charged with manslaughter, with a suspended jail sentence due only to his ill health. The company was fined 385,000 GBP, which was 250% of its turnover, and it inevitably went into liquidation. In another case in July 2011 involving a steel factory in Manchester, using the CMA-2007 and the HASWA-1975, three company directors were charged with gross negligence, manslaughter and failing to provide safe working conditions for their workers.[3]

There are two dichotomous approaches to the management of legal non-compliance, especially in occupational health and safety. The first is a school of thought that self-regulation is best, using legal punishment as a last resort. The second argues for more policing, enforcement and punishment in line with typical legal management of crime.[4] The self-regulation compliance scholars argue that those violating safety are different from common criminals because organizational employees (at all levels) engage with socially productive activities and therefore have the capacity to be socially responsible, unlike common criminals, who are less inclined towards socially responsible behaviour. The enforcement scholars argue that violations take place less due to incompetence and more to do with weighing up the impact of non-compliance against economic gains and even incompetence in the workforce. The debate continues, but in both models the onus for occupational health and safety falls on both employees and employers.

Another two recent significant developments have been: (1) "the internationalization of the law" – whereby even in the UK judges are starting to see that if international standards are grossly breached, there is greater room for court action; and (2) greater professional liability imposed on those who are advisors on risk such as EHS managers and consultants. Likewise, failing to act on recommendations from risk assessments or health and safety advice exposes organizations and managers to serious liability if something goes wrong.[5]

For example, the US's Federal Rule 404 explicitly excludes evidence of prior acts or occurrences to prove a person's (including a company's) character. In the case of the Deepwater Horizon oil spill, the prosecution wished to "argue that BP's past failures and its motives for failing to take costly steps to prevent the oil spill meant that the Deepwater Horizon Spill could not be considered an accident or mistake" (Brainich and Harris 2012, page 2). The court did not allow the evidence but left open the possibility that such evidence would be admissible at a later stage of the trial.[6]

In terms of the law, organizations must generally be able to demonstrate that they have taken the correct and reasonable steps to prevent incidents. Therefore they should conduct a risk assessment sufficient to the appropriate level required to address the risk (e.g. HSE-1996 – defining best practices etc.).[7] The lack of compliance even to basic safety requirements such as breach of fire escapes can bring both civil and criminal liabilities onto organizations.[8] For example, a survey of 164 fleet operators (employers) found 91% felt eye tests for their drivers were important, but only 38% actually had a policy on eye tests in place for their drivers.[9] Eye tests are not mandatory, but in terms of liability, an organization responsible for the transport of dangerous goods would be held liable for both civil and criminal charges if a driver employed by them was involved in a major accident leading to multiple deaths and injuries if they could not prove the driver's occupational fitness. Would failing to have a policy mean that the CEO/MD and board would then also be liable?

There are as yet few successful criminal prosecutions (in case law) of senior directors and CEOs or MDs, **but developments in the law show the need for them to have greater involvement in tackling organizational EHS risks**.

The question of how staff more generally can be held responsible for EHS risks is an important one. So although employers are the primary target of regulatory enforcement, in a neo-liberal sense health and safety responsibilities are equally shared between employers and their employees; but employees are in fact more responsible as they are closer to the risk. The "responsibilization" for safety by motivating workers to behave more safely and giving them greater empowerment to undertake tasks safely has corporate political problems associated with it. The trade unions have not favoured this as it puts greater onus on workers, alleviating the supervisors and other employer representatives from responsibility. In an immature organization and one where the workers are not knowledgeable enough, this may impact on productivity, and it has to be appreciated that for both workers and supervisors, compliance to safety norms and taking full responsibility is linked to a complex set of social and institutional relationships which are created through labour-market and workplace dynamics.[4]

For example, various laws in the Middle East such as the Labour Law No. 8 of 1980 and various ministerial decisions in the United Arab Emirates (UAE) may lead to prosecution of employers who fail to provide basic health and safety measures including preventative measures; first aid facilities and associated equipment; safe access and egress onto sites and industrial facilities; and suitable living accommodation, etc. The environment, health and safety laws in the UAE are more far-reaching than many other laws with regards to jurisdiction in the sense that no particular area is exempted (except perhaps the armed forces in the Ministry of Interior) and, as such, even Free Zones, which enjoy many exemptions from various regulations, must comply with HSE laws and regulations.[10]

In the Emirate of Abu Dhabi (the capital of the UAE), the establishment of the EHS Centre, created within the Environment Agency in 2009, passed a decree which launched the second version of the EHS Management System on 30 March 2012. This system, which covers all Sector Regulatory Authorities for Power and Water; Hospitality; Agriculture, etc. in Abu Dhabi, provides a common policy and manuals and unifies terminologies. The elements help in the establishment of different regulatory instruments such as codes of practice, standards and trigger values and mechanisms. These processes, starting with transparent and effective reporting on incidents, will help improve EHS performance.[11]

The state of Qatar published its National Development Strategy in March 2011, emphasizing the need for a robust health, safety and environment (HSE) regime. In the same year the National Committee on Occupational Health and Safety (OHS) was formed, which was to propose a national policy and system for OHS; devise and revise the OHS rules and regulations currently in force; and propose a mechanism for enforcing compliance. This led in May 2011 to a HSE legal framework document for the oil and gas sector which brought together all the rules and regulations currently promulgated.[12]

The Kingdom of Bahrain established the Supreme Health and Safety Council more than 15 years ago to develop and harmonize many safety rules, regulations and practices, with its members representing industry, the government and non-governmental organizations. Independent members were included for their wealth of knowledge and expertise.[13]

In the Sultanate of Oman, Kingdom of Saudi Arabia and State of Kuwait, many safety laws, rules and regulations have been developed, although implementation is complicated by many regulations falling under different jurisdictions. Given developments in Kuwait, Abu Dhabi and Dubai, consolidation of EHS rules and regulations is expected in the future.

The increase in regulations brings clarity to the judicial (and penal) system when allocating blame, especially when employers fail to fulfil their duties under these codes. This will lead to more change. A good example is the UAE Fire and Life Safety Code (2011), which was more or less ratified by all the GCC states who could adopt it; is very much an interesting development. To this end, EHS laws and regulations may not be driving a real and serious change in directors and CEO/MDs as yet, but with the increasing involvement of the public prosecutor's office and judges, the whole system's community is becoming more and more aware of the efforts that must be exerted to prevent incidents. This puts greater pressure on organizations and their leadership to establish preventative policies and strategies. Generally, one single piece of legislation does not cover all jurisdictions, but company duties and liabilities are often covered under common law, that is, judge-established case law.

Recent developments such as the Corporate Manslaughter Act (2009) in the UK are further developing liability issues and putting more emphasis on direct personnel liabilities of directors and managing directors.

Antrobus (2013)[14] gives a comprehensive review of the challenges in the implementation of the Corporate Manslaughter Act (2009) and explains that, in law, to charge a director under section 37 of the Health and Safety at Work Act (1975) and also Corporate Manslaughter Act (2009), the prosecution must prove that:

1 The defendant (director) owed a duty of care to the deceased;
2 The defendant breached that duty;
3 The breach caused the death.

In larger firms directors are generally not directly involved in the business, and proving those three points is challenging. Thus, most cases that have gone to trial so far have been of smaller firms with executive directors who are more involved in the day-to-day operations. It may be extremely difficult, therefore, to prosecute directors for gross negligence. Moreover, complying with regulations can be restrictive for a business, and there may be risks involved in what some regard as over-compliance. It is said that CEO/MDs themselves have expressed a wish to contribute by helping shape more risk-based or performance-based regulations.[15] This should be welcomed.

An independent review of the state of H&S legislation in the UK suggests there is sufficient regulation in place, and the challenge now is to enable businesses to reclaim ownership of the management of health and safety and see it as a vital part of the business rather than unnecessary bureaucracy.[16]

To conclude, whilst the law and legal system develop through time with more cases being tried, board directors must find a way of insulating themselves by expecting and monitoring their executive management teams to have the right systems in place. Whilst they continue to work on overall governance, they must educate and involve themselves somewhat with the developments through exploring the current and emerging risk borne from both their dynamic operations within organizations and the changing landscape of the regulatory regimes wherever they operate. This proves to be even more of a challenge for organizations that operate out of multiple jurisdictions and also to some extent those operating in emerging markets, where the laws and regulations are fast changing.

As such, even company secretaries, who are often legally trained individuals, should keep the chairman and the whole board abreast of health and safety laws that may impact their organization and their persons in terms of liabilities. This would include legal due diligences in new ventures, joint ventures and also mergers and acquisitions.

References

1 Institute of Directors (IOD) (1999): *Standards for the Board – Improving The Effectiveness of Your Board (Good Practice for Directors)*, Edited by Tony Renton. IoD and Kogan Page, London.
2 Metzger, Michael B. (Fall 1987): "Organizations and the Law", *American Business Law Journal*, 25(3), 408–441.
3 Forlin, Gerard (9th August 2011): "Developments in Health and Safety", *Archbold Review*, 7.
4 Gray, Garry C. (May 2009): "The Responsibilization Strategy of Health and Safety – Neo-Liberalism and the Reconfiguration of Individual Responsibility for Risk", *British Journal of Criminology*, 49(3), 326–342.
5 Forlin, Gerard and Smail, Louise (4th March 2011): "Come in Number 43", *New Law Journal*. Legal World, Law in Headlines, Legal Update 306, Specialist, available from www.newlawjournal.co.uk.
6 Brainich, Marc and Harris, Elliot (April 2012): "Significant Rulings From the Deep-Water Horizon Court on Discovery and Evidentiary Matters", Toxic Tort and Environmental Law Update, Sedgwick LLP.
7 Health & Safety Executive, "HSG-96" (1996): *Cost of Accidents at Work Guideline* (HSG-96). HSE Books, Her Majesties Stationery Office (HMSO), London.
8 Forlin, Gerard (23rd March 2012): "Too Hot to Handle – The Heat Is on for Organizations and Individuals Who Do Not Pay Heed to Fire Safety Precautions", *New Law Journal*, Legal Update 417, Specialist, available from www.newlawjournal.co.uk.
9 Roberts, Gareth (16th February 2012): "Inadequate Eye Tests Put Drivers' Lives at Risk", available from www.fleetnews.co.uk.
10 Kelly, Rebecca and Chicken, Laura (September 2011): "Health and Safety in the UAE – What Are Your Obligations as an Employer", *Insight* article, Clyde and Co.
11 Kelly, Rebecca and Chicken, Laura (2nd May 2012): "Environment, Health and Safety Legal Reform in Abu Dhabi", *Insight* article, Legal Updates, Clyde and Co.
12 Salt, David and Early, Michael (June 2011): "Health and Safety Developments in Qatar", *Insight* article, Clyde and Co.
13 Meeting with MoL in Bahrain, March 2013, Manama, Bahrain, in his office on the 13th March 2013 at 1000 Hrs, Note 1.
14 Antrobus, Simon (2013): "The Criminal Liability of Directors for Health and Safety Breaches and Manslaughter", *The Criminal Law Review*, 4, 309–322.
15 Richardson, Graham (2013): "Risk in Emerging Markets", *Governance*, March(225), 6–7.
16 Lofstede, Ragnar (November 2011): "Reclaiming Health and Safety for All: An Independent Review of Health and Safety Legislation", Presented to the Parliament by the Secretary of State for Work and Pensions by Command of Her Majesty, available from www.official-documents.gov.uk, printed by the UK Stationary Office Ltd.

5 Environment, health and safety (EHS) in organizations

Many practitioners and managers over the years have questioned if EHS management and management systems actually drive the development of a safety culture. If this were the case, then that would explain how organizations have changed from being high risk to high reliability industries in the past few decades.

There has been significant growth in interest in EHS in the past three to four decades, driven by various factors. These include the impact of legislation aimed to protect employees, contractors and the public from poor EHS practices.[1] For example, the Health and Safety Commission (HSC) in the UK adopted the recommendations of the Turnbull Report. Comprehensive advice was provided in "Implementing Turnbull".[2]

EHS is directly related to principles of loss prevention. Applicable in almost any business, it is perhaps more significant for the oil and gas and other high risk/high reliability industries where accidents can lead to considerable destruction to people and property as well as quickly erode share value, and as such EHS has become a significant business concern.[3] It must become a core personal value at the individual level if a safety culture is to be embedded.[4] It is important to appreciate that whilst financial losses can be insured to a great extent, other significant impacts on reputation, customer loyalty and stakeholder confidence (including public trust) can lead to considerable and irreparable damage.

Accidents cost companies money both directly and indirectly. The indirect costs of an incident can be estimated as being up to 30 times the direct losses caused.[5] Insurance may not cover lost production time, loss of highly trained personnel, impacts on employee morale and productivity and time and resources spent investigating the incident. The Health and Safety Executive (HSE) in the UK estimates that for every 1 pound sterling of insured loss there is an estimated uninsured loss of between 8 to 36 times more.[6]

This is exemplified in recent examples from the oil industry. The BP Texas Refinery incident in 2005 resulted in 15 fatalities and more than 170 injuries and cost BP both significant financial and reputational loss.[7] The 2010 disaster off the Gulf of Mexico in BP's offshore operations, which was one of the most serious in terms of impact on economy, the environment and people, led to the CEO's removal from his post for failing to demonstrate safety leadership.[8] In the past 5–10 years it is probable that no company has felt the crippling impact on its reputation (and shareholder confidence) and share price (company value) like BP since the Deepwater Horizon incident in late April 2010. The share price on 25 June 2010 (one week after the congressional hearing with BP's CEO) had dropped from 654.6 p to 304.6 p (i.e. it lost about 46.5% of its original value). Even on 20 October 2011 (more than 18 months on) the share price remained just over 460 p.[9]

But, as Haefeli et al.[10] (2005, page 5) explain:

> Most organizations were concerned about potential cost implications of major incidents, but were less concerned about actual costs incurred as a result of more frequent, minor events. The majority of respondents reported that they did not know how much either accidents or work related illnesses were costing their business. Few attempts had been made to quantify the cost of health and safety failures. Limited time and resources, perceived complexity and lack of expertise were the most commonly cited barriers to conducting accident/work-related ill health cost assessments.

This is an important finding given the extensive research that statistically links near-miss and minor incidents with major incidents including fatalities. The US Labour Force Survey in 1990 established ratios relating minor incidents to lost-time (more than three days off work) and major incidents. In the UK the RIDDOR regulation links lost-time incidents to fatalities with a ratio of 400:1. Earlier, Frank Bird established a ratio of 600:1 in terms of near misses to major incidents. Heinrich's Domino Theory, established in the 1960s, explains that an incident is caused by a failing of barriers to control or eliminate unsafe conditions and acts. If these persist and thus near-miss incidents occur, it is statistically significant that at some point a major incident will occur. Theory places cultural aspects and social environmental factors as root causes, thus including recklessness, stubbornness and greed.[11]

This would mean that although senior managers interviewed in one study[12] were concerned about the major accidents, they may not have realized that controlling or reducing minor incidents prevents the major incidents they were concerned about and that they need to focus on achieving "behavioural changes among staff at lower levels within organizations, as well as tapping into the moral obligations of senior managers and boards of directors" (Haefeli et al. 2005, page 170).[10] This is difficult given incorrect reporting lines of HSE practitioners and the lack of appropriately competent and trained staff.[7]

Many of the major investigation reports into some of the most significant recent accidents, such as the BP Texas refinery explosion in 2005,[13] the Piper Alpha Incident in 1988[14, 15] and the explosion/fire at Buncefield Oil Terminal in 2005,[16, 17] have emphasized the failure of management more broadly and the company leadership most particularly in preventing such incidents.

As explained earlier, accidents cost companies money both directly and indirectly, and this is essential to understand as a CEO or board director. Certain types of insurance protect employers, and these include employer's liability, public liability, workman's compensation, fire and perils and so on. It is to be noted that losses cannot always be recovered for matters such as lost production time, loss of highly trained personnel, impacts on employee morale and productivity and time and resources spent investigating the incident etc.

So what is this safety culture which needs to be developed and driven from the top of the organization? Past disasters have shown the need for a real and strong commitment from the corporate and senior management.[18] Thus the Bahraini Petroleum Oil Company (BAPCO) developed effective risk assessment (RA) and quantitative RA (QRA) programmes after an in-depth review and investigation of the Texas Refinery incident of March 2005. Driven from the top, changes include using Port-a-cabins that are blast proof.[19] One of Saudi Aramco's affiliates changed its focus to make leadership and accountability the most important element of the company's safety management system.[20] DuPont's PSM system considers EHS a business issue – not an operational and manufacturing issue – to ensure management commitment to uncertainty avoidance. Dupont developed a global contract management system which includes six elements:

1   Contractor selection;
2   Contract preparation;
3   Contract award/establish expectations and standards;
4   Orientation and training;
5   Monitoring safety activities; and
6   Evaluating safety performance against contractual expectations.[21]

A just culture is the foundation of any effective safety culture.[22] Error Management and Total Error Reduction Management (TERM) systems are very effective tools for managing incidents by identifying a series of contributing factors for an incident – i.e. a collection of causes. A "just culture" allows for the reporting of incidents openly and reduces the number of accidents by limiting the incidents through effective prevention by not penalizing the reporting party or otherwise. Reporting near misses can help identify where the next incident will most probably occur. There is a great misunderstanding of near misses; it is about organizational culture – **management must follow up positively and see how things are being addressed.** High potential near-miss incidents (HPNMI) should be investigated in the same way as actual incidents leading to serious damage and loss.[23]

Near misses are ultimately a great opportunity to learn for an organization and very specific to what is happening on that site – although a blame culture can inhibit this.[24] Further, in the Middle East most PSM incidents are caused by contractors, who have workforces comprising many different nationalities and languages. Gaining compliance of contractors in training their workforce effectively and in monitoring and documenting, etc., is therefore challenging.[24]

One major issue is that contractors with substandard safety performance may be appointed to a project based on cost considerations. In almost all organizations such large contracts *require the review/approval of the board or at least investment committees with BoD members.* It is in these "due-diligence" forums that safety performance and standards must be challenged. This explains Dolphin Energy's management system emphasis during pre-qualification of contractors, as elaborated by Al-Rahbi (2008),[25] where in contractor questionnaires and pre-qualifications, two of the key elements out of 12 are contractor's management commitment to safety and HSE aspects and also the allocation of resources and organization to projects.

Senior managers must appreciate that the driving motivation for contractors is profitability. In the GCC contractors are driven mostly by price, and the risks of operating within a live plant can be exponential. Unless clients/employers set a higher standard, contractors will continue to be the biggest and weakest link.[26] Some are attempting to tackle this: Hemler (2010)[27] explains that Saudi Aramco's Contractor Safety Management System requires contractors to establish a programme to establish accountability, communications requirements, performance measures and standard maintenance through compliance and monitoring activities. He emphasizes that this can only start with effective pre-qualification; pre-job safety discussions; facility safety orientation; and site safety performance monitoring. **He goes on to explain that none of this can be truly implemented without top management commitment.**

The relationship between safety and leadership is discussed further in this book. But a good safety culture clearly requires a just and fair organizational culture; strong management commitment and leadership with strong governance systems; a clear understanding that eliminating major incidents starts with eliminating the smaller incidents and near misses; and a strong focus on the weakest links such as contractors.

And whilst safety management is a line manager's ultimate responsibility, this in no way absolves the senior leadership team's accountability for oversight, performance evaluation and ensuring that effective risk management is in place and the operations are being managed by competent, trained and responsible individuals working within a robust framework and management system.

References

1 Health & Safety Executive (2001): *A Guide to Measuring Health and Safety Performance.* HSE Books, Sudbury, Suffolk, UK.

2 Institute of Chartered Accountants (UK) (September 1999): "Implementing Turnbull – A Boardroom Briefing", Centre for Business Performance Thought Leadership From the Institute, The Institute of Chartered Accountants in England and Wales, Accountancy Books (Ref 2).

3 MacLean, Richard (2007): "Environmental Leadership – Get Organized", *Environmental Quality Management Journal*, Winter, 95–98.

4 Al Hamoud, Ali M. (February 2010): "How Culture Influences Behaviour Based Safety in the Workplace", ASSE-MEC-2010-04, American Society of Safety Engineers – Middle East Chapter Conference and Exhibition, Bahrain, Page 64.

5 DNV (1996): "Loss Control Management", Modern Safety Management Training Program: Module on the Causes, Effects and Control of Loss.

6 Health & Safety Executive, "HSG-96" (1996): *Cost of Accidents at Work Guideline* (HSG-96). HSE Books, Her Majesty's Stationary Office (HMSO), London.

7 Baker, James (III), Bowman, Frank L., Glen, Erwin, Gorton, Slade, Hundershot, Dennis, Levison, Nancy, Priest, Sharon, Rosentel, Tebo, Paul, Wiegmann, Douglas and Wilson, Ducan (2007 January): "The Report of the B.P. U.S. Refineries Independent Safety Review Panel", available from www.csb.gov.

8 Al Hashmi, Waddah Ghanim (2011): "Safety Leadership Case Study – An In-depth Analysis of the Congressional Hearing with Tony Hayward, BP ex-CEO on the Deep-Water Horizon BP Oil Disaster, 2009 in the Gulf of Mexico, USA", Working Research Paper 01/11, Presented at 9th Annual OHS Congress, 5th–10th Feb 2011, Dubai, UAE.

9 London South East website, www.lse.co.uk, accessed 30th June 2014.

10 Haefeli, Karen, Haslam, Cheryl and Haslam, Roger (2005): "Perceptions of the Cost Implications of Health and Safety Failures", A Research Report Prepared for the Health and Safety Executive (2005) by the Institute of Work, Health and Organizations and the Health and Safety Ergonomics Unit, Research Report No. 403, Published by the Health and Safety Executive (HSE) Books.

11 British Safety Council (BSC) (2006): "International Diploma in Occupational Safety and Health, Module A", Sub-Element British Safety Council, Issue 2, January 2006, British Safety Council Awards.

12 MacLean, Richard (November 2011): "Size Matters", *Environmental Matters (EM), Air and Waste Management Organization Magazine*, 52–53, available from www.awma.org.

13 Kumar, Hari (December 2007): "Piper Alpha Disaster – An Overview", Presentation at the EHSSQ Workshop for the Horizon Terminals Group", Dubai, United Arab Emirates.

14 Cullen, Douglas W. (Lord) (1990): "The Public Enquiry into the Piper-Alpha Disaster", Her Majesty's Stationary Office (HMSO), London.

15 Allars, Kevin (UK-HSE) (2007): "Buncefield, the Story so Far – Case Study Presentation", Delivered at the Platts: Creating Value in Oil Storage, 26th–27th November 2007, Budapest, Hungary.

16 Major Incident Investigation Board (MIIB) (2011): "The Buncefield Incident 11 December 2005: The Final Report of the Major Incident Investigation Board", Control of Major Accident Hazards (COMAH), available from http://.buncefieldinvestigation.gov.uk.

17 Patel, Jitu (February 2012): "Safety Culture Behaviour – A Global Challenge 'Revolutionizing Safety Management System'", ASSE-MEC-2012–20, American Society of Safety Engineers – Middle East Chapter Conference and Exhibition, Bahrain, Page 146.

18 Goyal, Ram and Menon, Vinod (February 2012): "Location of Permanent and Temporary Building in Process Plant Areas – Practical implications of the API Recommended Practices", ASSE-MEC-2012-63, American Society of Safety Engineers – Middle East Chapter Conference and Exhibition, Bahrain, Page 437.

19 Al-Kudmani, Ahmed (February 2008): "Building a Safety Culture – Our Experience", ASSE-MEC-0208-21, American Society of Safety Engineers – Middle East Chapter Conference and Exhibition, Bahrain, Page 137.

20 Van der Westhuyzen (2), Johan (February 2012): "Ensuring Contractor Alignment With Safety Culture", ASSE-MEC-2012-52, American Society of Safety Engineers – Middle East Chapter Conference and Exhibition, Bahrain, Page 372.

21 Bu-Allay, Khalid (February 2012): "Understanding, Identifying and Managing Maintenance Errors in Military Aviation", ASSE-MEC-2012-51, American Society of Safety Engineers – Middle East Chapter Conference and Exhibition, Bahrain, Page 351.

22 Vasudeven, N. and Dutta, Dinish Kumar (February 2010): "New Approach to Enhance Near Miss Incident", ASSE-MEC-2010-51, American Society of Safety Engineers – Middle East Chapter Conference and Exhibition, Bahrain, Page 378.

23 Basson, Wayne (February 2010): "Negligent Employees . . . A Personal Journey Across 5 Continents", ASSE-MEC-2010-49, American Society of Safety Engineers – Middle East Chapter Conference and Exhibition, Bahrain, Page 365.

24 Snakard, Mike and Hazzan, Mike (February 2010): "Lessons Learned in Managing Contractor Safety", ASSE-MEC-2010-21, American Society of Safety Engineers – Middle East Chapter Conference and Exhibition, Bahrain, Page 175.

25 Al-Rahbi, Ismail Ali Khalifa (February 2008): "HSE Management for Contractors in the Oil and Gas Industry", ASSE-MEC-0208-07, American Society of Safety Engineers – Middle East Chapter Conference and Exhibition, Bahrain, Page 47.

26 Drelaud, Jason (February 2010): "Contractors – A Necessary Risk", ASSE-MEC-2010-20, American Society of Safety Engineers – Middle East Chapter Conference and Exhibition, Bahrain, Page 171.

27 Hemler, Steven (February 2010): "Implementing a Comprehensive Contractor Safety Management Program", ASSE-MEC-2010-19, American Society of Safety Engineers – Middle East Chapter Conference and Exhibition, Bahrain, Page 166.

6 Risk perception, risk management and risk tolerance

In this book probably one of the most significant areas to be explored is risk management. The literature has a significant abundance of studies and articles that talk about risk and the role of risk management in organizations. The definition of risk and risk control is important to establish. Equally important is the understanding and appreciation of how risk perception at the senior management and leadership levels can impact on the type, quality and speed of decision making in high risk/high reliability organizations. The concept of loss prevention and loss control in the context of the wider enterprise risk management (ERM) will be explored. The knowledge and appreciation of risk is so fundamentally important for the CEO/MD and the board to understand that it can be described as the single most significant competency.

The management of risk concerns itself with the prevention of loss, preventing being negatively impacted by issuance of enforcement, improvement or prohibition notices from the "local authority having jurisdiction" for health and safety. It also helps avoid punitive action in which both civil and criminal courts may impose fines and compensation claims and imprisonment for breaches of legal duties, respectively, and which can affect companies or individuals and thus their operations and reputation.[1] Riaz-ul-Hassan (2012) makes reference to the Deepwater case study and says: "Better management of decision-making processes within BP and other companies, better communication within and between BP and its contractors and effective training of key engineering and rig personnel would have prevented the Macondo incident" (Riaz-ul-Hassan 2012, page 83).[2]

A cost value analysis of safety incidents must be studied by the senior management. This was based on a study that was undertaken under board guidance for a high risk chemical company which used quantitative loss evaluation methods to manage risks.[3]

Mandel (2012) talks of the stages of the development of management of risk in organizations and argues that over time the value of risk management has been mainly driven by changing the business needs perspective from financial to operational to management to strategic. He offers an explanation which has been summarized in Table 6.1.[4]

Table 6.1 Corporate Enterprise Risk Management

Perspective	Corporate Value	Focus	Scope	Type of Risk Management
Financial	Low	Hazard and casualty risks	Risk transfer, insurance, loss prevention or mitigation of insured risks	Defensive
Operational Management	Medium	Individual business risks	Mitigation of controllable risks and management of risk as an expense	Advanced
Strategic Management	High	Strategic and operational risks	Support of business objectives, consistent, systematic risk management practices and risk as a differentiator	Enterprise

Risk management is one of the primary responsibilities of directors, and as such they are required to provide leadership within the framework of prudent and effective controls which enable risk to be assessed and managed. The Combined Code states:

> Are the significant internal and external operational, financial, compliance and other risks identified and assessed on an on-going basis? (Significant risks may, for example, include those related to market, credit, liquidity, technological, legal, health, safety and environmental, reputation, and business probity issues).
>
> (HSE 2006, page 10)[5]

Leadership includes setting the strategic direction; setting values and standards of business conduct and objectives; holding management accountable for actions; upholding obligations to all stakeholders; and overseeing the internal controls and assessing their effectiveness.

To define risk first in some context, Mandel (2012) defines strategic risk as "those internal or external uncertainties, whether event or trend driven, which impact an organization's strategies and or the implementation of its strategies" (Mandel 2012, page 11).[4] An example of dealing with external risks would be if security risk to installations is considered; in general high risk operations are well security managed to ensure they control any kinds of imported risks. The issue of the changing face of security is becoming more integrated as part of the installation enterprise safety management system and is also being demanded by many insurers of risk, who end up carrying the burden of *organizationally transferred* risk.[6]

There are some quantitative studies that have showed through using a multivariable frequency analysis of different HSE incidents that cost control can be achieved by better HSE risk management; thus good safety equals good business.[7]

In reference to one of the widely used risk management (RM) standards across the industry today, a structured process to manage risks arising from operations including environment, health, safety, quality and security, and financial as well as reputational is imperative for organizations. He goes on to explain the ISO 31000 International RM Guidelines and tries to address the question: "Is the level of risk acceptable, and does it require further considerations and actions?" This raises issues relating to uncertainty and risk appetite. He also explains that RM must be integrated within the organizational management systems and built on continual improvement and dynamic reviews.[8] Clarke (2012), on the other hand, explained in detail the concept of "As low as reasonably practicable – or ALARP" as the way that many organizations today deal with rationalizing their risk appetite decisions.[9]

There is a compelling case for risk governance at the board level, as directors have to be concerned about risk, which includes the fiduciary duties; how risk contributes to an organization's strategy; constructively challenging management's proposals; due diligence and risk awareness; and so on. There are five key oversight responsibilities which are common to many of the leading corporate governance codes.[10]

These are summarized in Table 6.2:

Table 6.2 Summary of Directors' Risk Oversight Responsibilities

No	Theme	Action
1	Strategy	Approve strategic planning processes and the organization's strategic plan, which includes sustainability of operations, opportunities and risks.
2	Risk management	Review and approve the main risks associated with the organization's activities.
3	Compliance	Ensure that processes and systems exist and are being implemented to manage those risks – including systems of internal control.
4	Processes	Define at a high level and approve processes in which the BoD or one of its committees evaluates the company's main risks (periodically).
5	Structures	Review and approve the organization's structures and processes to manage both existing and emerging risks.

Table 6.3 Risk Management – Emerging Best Practices

No.	Best Practice	Action Description
1	Risk appetite	Develop a risk appetite statement (developed by the executive management and approved by the board) that expresses the attitude of the organization towards risk taking, at times setting lower and upper limits.
2	Risk culture	Studies have shown that catastrophic losses have arisen from a lack of risk management, a culture which pays little attention and tolerates, or worse, encourages risk-taking behaviour. A culture of risk management needs to be set; this has to be set by both the executive and BoD, and this needs to percolate throughout the organization.
3	Risk committee	Many best-practice codes have required that boards set up audit (and risk) committees. Risk committees looking more holistically at internal and external risks and involving a wider base of professionals is where newer risk management guidelines are moving.
4	Chief risk officer	Especially in large and more complex organizations, the role of a chief risk officer who reports directly to the CEO and has access to the board is where larger, more progressive and higher reliability organizations have been moving.
5	Internal audit function	A separation of the risk and audit functions has become more accepted. Even between these two functions, due to the inevitable internal/external focus which develops, a possible conflict of interest may arise, and thus these functions in more complex and progressive organizations have been physically separated.

Trends show that boards are moving toward the following best practices when it comes to risk management (Dufort 2013)[2] (see Table 6.3).[11]

O'Conner (2010) defines ERM as:

> A Process, affected by an entity's Board of Directors, Management and other personnel, applied in strategy setting across the enterprise, designed to identify potential events that may affect the entity, and manage risks to be within its risk appetite, to provide reasonable assurance regarding the achievement of entity objectives.
>
> (O'Conner 2010, page 46)[12]

In another definition by Deloitte (2008, page 12):[13]

> ERM is an enabler of risk intelligence, and its true value may lie in its ability to enable a systematic identification of possible causes of failure – failure to protect existing assets and failure to achieve future growth, i.e. manage both rewarded and unrewarded risk. Unrewarded risks are typically associated with lack of integrity in financial reporting, non-compliance with laws and regulations, and operational failures – i.e. there is no premium to be obtained for taking these types of risks. Rewarded risks are those that typically have to do with strategy and its execution.

Whilst this may be true, it must also be equally appreciated that it is not always easy to differentiate between them, as they are at times interconnected. To give an example, an overall strategy to reduce cost of maintenance may lead to a rapid deterioration of integrity and thus lead to a serious incident that cripples an operation.

Between 2008 and 2010 economies were in the midst of a global financial crisis; organizations started looking more closely at enterprise risk management.[14] In fact a study undertaken by Deliotte stated that the top risks facing the energy sector (i.e. power and utility, system operators, and oil and gas) in the Middle East included *regulatory risks; asset performance (integrity); operating integrity; business continuity; and people/talent*. Whilst these are not EHS risk directly, they can have both a direct and indirect impact on EHS.[15]

The Jaipur October 2009 fire incident in India killed 11 people, injured another 45, destroyed the oil storage terminal and damaged properties up to 2 km away. The incident was a result of a routine operation. Six major recommendations to the management committee of the Indian Oil Corporation were made, which included **in-depth board-level reviews on a quarterly basis focusing on safety in the different sections within the organization, with a primary focus on risks and mitigation measures**.

It also was recommended that the executive management at the group level undertake at least two inspections of every major installation within their areas of responsibility per annum to look at risk control measures and emergency response preparedness and report this back to the CEO. Major safety reviews and audit findings must be personally reviewed by the CEO.[16]

It is also recommended that higher management's outlook about managing plant life-cycle risk must integrate with both the EHS risk profile and the organization's investment portfolio and must address the supply chain impacts, running scenarios and looking at the risk more holistically as a model rather than just the plant risk.[17]

Risk can be extremely subjective and must be considered in a wider context of environment, people, organization and cost-benefit analysis.[18] In a case study of upstream O&G operations, the HSE philosophy was built on various drivers, of which the first is to identify the high risk areas. The main driver for the executive management team was to bring about a consistent review process focusing on the high risk areas in a performance-based approach.[19]

McKinnon (2012) explains that in the domino theory, the causality between underlying factors and the sequence of events is governed by good management system practices to control loss.[20] However, Bibbings (2001) maintains that the future of boards will mean expectation for greater competency to give stakeholders "Control Assurance".[21]

A Global Risks 2012 report issued by the WEF's Insight report concluded that leadership at both a national and organizational level needs to improve the real and perceived risks within the industries with respect to public safety and in the tools of communication. Another critical recommendation is that transparency needs to improve in such a way that sharing information on risks must be improved, to in turn improve public perception of the risk – i.e. give them better understanding. Many of the risks discussed in the report are macro such as the impact of global climatic changes etc. Whilst these are very high level risks and little can be done at an organizational level, whatever little can be done should be done, and at the very least directors and executives of organizations need to understand them better in this ever-faster globalizing world.[22]

Deloitte[23] (2008) defines six major areas of risk management which must be considered by major organizations, which include:

1   Defining the board's risk overview;
2   Fostering a risk-intelligent culture;
3   Integration and incorporation of risk intelligence into strategy;
4   Defining the risk appetite;
5   Executing the risk intelligence governance processes; and
6   Benchmarking and evaluating the governance processes.

They argue that these six areas of focus would reflect the view that risk taking for reward and growth is as important as risk mitigation to protect current assets and operations. Once this becomes part of the considerations always looked at in the decision-making processes by boards, generally a more risk-averse strategy develops. This in turn will effectively ensure more of a sustainable and robust business model for any high risk organization following that; as such this would effectively build reliability into the strategy.

Because more complex organizations have interconnected processes, whilst an operational risk may be credible and accepted by a board, such as a fire in a plant, the magnitude of the impact of the direct damages is by far less than the implications of the stoppage of operations and the repercussions on the supply chain and business continuity. The Buncefield Oil Storage facility (UK) is a great example, where in 2005 it was completely incapacitated after an explosion/fire and which took seven days to control, with immense environmental damage. The terminal supplied 25–30% of jet fuel to Heathrow Airport (the busiest international airport in the world), and when those supplies were interrupted, there were serious operational and financial implications to Heathrow.[23]

With reference to survey results of national oil companies (NOC), one of the top three risks along with political instability and rising operating costs is the environmental (EHS) risks and concerns, according to NOCs, international oil companies (IOCs) and independents. The oil service companies also see EHS as a top-three risk.[24]

So to conclude, whilst risk assessments may become very detailed and technical and require a good understanding of the company's operations, board directors must engage with the CEO/MD and their executive management in productive and meaningful discussion around risk management. It is essential, as they are the ultimate risk governors and the guardians of the company's health. Usually a good understanding of the organization's effective enterprise risk management system (ERM) is quite essential.

References

1 Lukic, Dane, Margaryan, Anoush and Littlejohn, Allison (2010): "How Organizations Learn From Safety Incidents: A Multifaceted Problem", *The Journal of Workplace Learning*, 22(7), 428–450.

2 Riaz-ul-Hassan (February 2012): "Human Factors Contribution to Risk (Human Factors– Reliability & PSM)", ASSE-MEC-2012-09, American Society of Safety Engineers – Middle East Chapter Conference and Exhibition, Bahrain, Page 82.

3 Bardy, Mariana, Cavanagh, Nic, Oliveira, Liz Fernando (February 2008): "Managing Business Risks From Major Chemical Process Accidents", ASSE-MEC-0208-35, American Society of Safety Engineers – Middle East Chapter Conference and Exhibition, Bahrain, Page 238.

4 Mandel, Chris (2012): "Integrating Enterprise Risk Management and Strategic Planning Workshop, May 29–30th 2012, Dubai, UAE", 2nd ERM Conference 2012, RIMS, Risk & Insurance Management Society.

5 Health & Safety Executive (HSE) (2006): "Defining Best Practice in Corporate Occupational Health and Safety Governance", A Report Prepared by Acona Ltd for the Health and Safety Executive Research Report 506.

6 Gregory, Stephen (2011): "Risk Management and Insurance: From the Guardroom to the Boardroom", Financier Worldwide – Reprinted From Global Reference Guide – Risk Management and Insurance.

7 Cavanagh, Nic, Hickly, Colin and Linn, Jeremy (February 2008): "Integrating Risk into Your Plant Lifecycle – A Next Generation Software Architecture for Risk Based Operations", ASSE-MEC-0208-47, American Society of Safety Engineers – Middle East Chapter Conference and Exhibition, Bahrain, Page 317 (Ref 2).

8 Newberry, Jim (February 2012): "The Gold Standard for Risk Management – Get a Grip on ISO 31000", ASSE-MEC-2012-07, American Society of Safety Engineers – Middle East Chapter Conference and Exhibition, Bahrain, Page 68.

9 Clarke, Sharon (2012): "Safety Leadership: A Meta-Analytic Review of Transformational and Transactional Leadership Styles as Antecedents of Safety Behaviours", *Journal of Occupational and Organizational Psychology* 83(1), 22.

10 Dunfort, Ghislain Giroux (2013): "The Risk Governance Imperative – Part 1", *Governance*, March(225).

11 Dunfort, Ghislain Giroux (2013): "The Risk Governance Imperative – Part 2", *Governance*, April(226).

12 O'Connor, Troy (February 2010): "Enterprise Risk Management – The Impact on SHE", ASSE-MEC-2010-02, American Society of Safety Engineers – Middle East Chapter Conference and Exhibition, Bahrain, Page 46 (Ref 1).

13 Deloitte (2008): "Perspectives on ERM and the Risk Intelligent Enterprise", The Enterprise Risk Management Benchmark Survey, Deloitte Development LLC.

14 Yousef, Deena Kamel (27th June 2010): "Energy Sector Deals With Risk", *Gulf News* Newspaper Article.

15 IOC – Indian Oil Corporation (2010): "IOC Fire Investigation Report in Jaipur, India on 29th October 2009", Section 10.15, Page 164, 2010 – Downloaded as part of the Shell Learning From Incidents Awareness Alert 2010/01 Report, available from www.oisd.gov.in/INDEPENDANTCOMMITTEEREPORT.HTML, accessed 1st September 2012.

16 Cavanagh, Nic, Hickly, Colin and Linn, Jeremy (February 2008): "Integrating Risk into Your Plant Lifecycle – A Next Generation Software Architecture for Risk Based Operations", ASSE-MEC-0208-47, American Society of Safety Engineers – Middle East Chapter Conference and Exhibition, Bahrain, Page 317 (Ref 2).

17 Iskandar, Mann M. (February 2010): "Behaviour and Human Factors in Safety", ASSE-MEC-2010-30, American Society of Safety Engineers – Middle East Chapter Conference and Exhibition, Bahrain, Page 242.

18 Al Nakib, Mohammed A. and Jackson, Chris (February 2008): "HSE Philosophies Innovative Approach to Project Design", ASSE-MEC-0208-41, American Society of Safety Engineers – Middle East Chapter Conference and Exhibition, Bahrain, Page 287.

19 McKinnon, Ron C. (February 2012): "Do We Measure Luck or Control", ASSE-MEC-2012-40, American Society of Safety Engineers – Middle East Chapter Conference and Exhibition, Bahrain, Page 281.

20 Bibbings, Roger (June 2001): "Corporate Risk Management", *Parting Shots*. Royal Society of Prevention of Accidents (RoSPA), Pages 22–24.

21 WEF, World Economic Forum (2012): "Insight Report – Global Risks 2012 – An Initiative of the Risk Response Network", World Economic Forum, Seventh Edition.

22 Deloitte (2008): "Perspectives on ERM and the Risk Intelligent Enterprise", The Enterprise Risk Management Benchmark Survey, Deloitte Development LLC.

23 Major Incident Investigation Board (MIIB) (2011): "The Buncefield Incident 11 December 2005: The Final Report of the Major Incident Investigation Board", Published by the Control of Major Accident Hazards (COMAH), available from http://.buncefieldinvestigation.gov.uk.

24 O'Dwyer, Terence (2013): "Risks, Opportunities and Challenges for National Oil Companies – Survey Results", World National Oil Companies Congress, London, 19–20 June 2013.

7 Relationship of corporate social responsibility (CSR), social accountability (SA), and sustainability

Corporate social responsibility (CSR), social accountability (SA) and sustainability all have a bearing on the integration of management systems and the positioning of organizations in the market and are interlinked in certain elements to EHS.[1] With CSR, the community work that organizations do to improve the safety and security of people's lives within regions in which they operate is important. HSE matters are key business concerns which impact on costs and give a competitive edge – *risk management as they see it sits on a foundation of commitment from the organization's corporate leaders.*[2]

With social accountability standards, safety and security elements are fundamental in the very basic standards of the International Labour Organisation (ILO). Sustainability reporting relates to environmental protection and the safety of the greater environment and that of employees and the public alike.

In the implementation of CSR and EHS systems, getting employees engaged is very important. Employee engagement can lead to very positive organizational and business benefits such as lower absenteeism (injury/illness rates are lower). It should be noted that more than 50% of injuries that employees have are off work; thus engaging them and even their families in safety has immense benefits to the organization as well as society at large.[3]

Voluntary reporting has increased in present years as organizations want to present themselves as good corporate citizens, and charity must start at home – when protecting their own employees. There is both academic and practice-based literature in these areas which addresses the involvement of leadership and company boards in driving these initiatives, endorsing reporting and enhancing transparency within their organizations, the industry and the general public. This is also becoming very important in the rapidly transforming context of the new socio-economic realities in the Middle East.

The Safety and Health Sustainability Taskforce set up by the American Society of Safety Engineers (ASSE)[4] had developed a Safety and Health Sustainability Index (SHSI). This index was built on six key elements: **Values:** (1) Safety and Health Responsibility Commitment; (2) Codes of Business Conduct; Operational Excellence; (3) Integrated and Effective Safety and Health Management System; (4) Professional Safety and Health Competencies; and under **Oversight and Transparency**: (5) Senior Leadership Oversight and Safety and Health; and (6) Transparent Reporting of Key Safety and Health Performance Indicators.

The inception of the ISO 9001 Quality Management System Standard in the early 1990s (which started as the British Standard, BS 5750) followed by the ISO 14001 Environmental Management System (EMS) and later the OHSAS 18001 Health & Safety Management System Standards have all brought about change in organizational behaviour towards self-driven compliance. These certifications, it may be argued, have given organizations an effective brand-value proposition and marketing edge against their competitors – with their stakeholders, more inclusively, rather than just their shareholders. This perhaps reflects the appetite to invest and comply with a standard when an organization feels it adds value from an external perspective.

Most recently the new ISO 45001 Standard, which is the new Health and Safety Management System, is a step towards making management more committed and personally engaged. The standard, which has been delayed with many consultation rounds and will be officially issued in the middle of 2018, places a greater emphasis on leadership engagement. It also expects that the context of the organization is better expressed. This also alludes to the organizational values being expressed more overtly.

The "Rewarding Virtue" document[5] recommended six areas in order to reinforce the UK's Combined Code. These include (1) setting of clear values and standards by the leadership; (2) thinking strategically about corporate responsibility; (3) being constructive about regulation by being self-regulating and supporting the authorities; (4) aligning performance management systems to encourage rewarding a longer-term outlook/behaviours rather than shorter-term and narrow financial targets; (5) creation of a culture of fairness and integrity in which the tone is set right at the top; and finally (6) using internal controls to secure responsibility through effective governance systems.[6]

A shift has taken place with organizations in the past 50–60 years from an obligation to a strategy when it comes to social value proposition. The links between the profit-making organizations and the more philanthropic ones have matured and emerged to become more symbiotic, supporting the greater development of resources such as marketing, technical and employee volunteerism. This meant more personal involvement of the organization's staff, with support from their employers rather than just paying into NGOs' cash contributions.[7]

Good companies continue to fail to do what is perceived to be the right thing. They fail to be able to clearly prevent things happening and things or situations deteriorating, and Schwartz and Gibb (1999)[8] conclude in their book, *When Good Companies Do Bad Things*", the following reasons why companies fail:

1   They fail to create a culture that tolerates dissent or one in which the planning processes are encouraged to take non-financial risks seriously;
2   They focus primarily on financial performance;
3   They do not positively support their employees in thinking about their work as a cohesive part of a team, working and using their moral and social intelligence as well as their business intelligence;

4 They focus on people and organizations that think and behave the same way and avoid those who do not agree with them or criticize them;

5 They let their commitments to certain projects and products overwhelm all other considerations and decisions, be they financial, ethical or social etc.; and

6 The senior management consider social issues as those for others to have to worry about, as this is not really part of their necessary operability and existence.

The notion that such companies do not really have a long-term view or vision in a social context and that they expose themselves to more bad incidents occurring is evident, first because their risk assessments are flawed and second because when there is a failure they have very little to show for having done anything to have effectively prevented it. As they must invest in emergency and crisis management, they become classified as *highly unreliable organizations.*

Another aspect that should be appreciated is that organizations behave differently in different parts of the world. The concept of "doing good to look good" is at odds with the more prevalent culture in the GCC and Middle East, which is more rooted in a tradition dating back to the Prophet Muhammad's (PBUH) conduct, metaphorically describing the fact that the right hand should not know (what charity) the left hand has given. This means that it is quite a foreign concept in the Arab/Muslim world to over-advertise this kind of CSR/charitable work. The concept is of course that the reward is in the hereafter. This is obviously not limited to the Arab/Muslim cultures.

In terms of environmental protection, social responsibility and EHS at work, these aspects have become of significant importance to corporations. Many organizations within the oil and gas sector, for example, will be very clear and vocal in their commitment to these issues. Trust in an organization by all its stakeholders, including the employees, customers and the public, is essential for its longevity and sustainable existence and growth.[9] This has led to the development of various committees, codes of ethics and CSR-type policies etc. The value of corporate governance goes beyond control in that it creates an environment of enterprise and best professional practice to extract the *long term-value from a commercial enterprise.*[10]

In 1997 a standard was issued (later updated in 2001 with a latest revision in 2014) as a guide to companies in addressing worker rights. This was the Social Accountability Standard SA 8000, which was developed by Social Accountability International, based in New York.[11] However, whilst the standard is novel, as it easily addresses the social accountability needs of organizations, it really infuses the main norms of the ILO's conventions relating to the Universal Declaration of Human Rights and the UN Convention of the Rights of a Child. The standard looks at issues from child labour, to forced labour, to freedom of association and the right to collective bargaining as well as EHS and working hours. It remains a voluntary standard but has driven many large organizations which operate and engage with businesses in third world countries to get pre-qualified and continually compliant with certain basic SA standards. This prevents organizations being blamed for exploitation and/ or even subsequently boycotted. Epstein (2008),[12] explains that whilst the Global Compact has helped in shaping human rights expectations of employers, it has had its fair share of criticism due to the failing or lack of monitoring, accountability and enforcement. Perhaps one of the reasons was that much of this drive has been overwhelming for organizations who wanted to comply, as they understood the importance, but in all fairness perhaps did not expend enough effort in initiating/inducting (also sometimes called "on-boarding") effectively all the leadership teams within these organizations, starting with the board of directors.

When organizations address the issue of sustainability, it is critical to understand three key factors: (1) greater environmental awareness in the public; (2) greater expectation from the shareholder for the board and management of an organization to ensure long-term (sustained) value proposition and; (3) the significantly increased "customer power" in that the customer has a greater choice to go to the extent of boycotting a product or service.

Bell and Morse (2008)[13] explain that "Greening the Strategy" is essential for many organizations today. This means taking steps including risk reduction and reducing environmental stresses and in turn the human vulnerability to environmental stress, and, in fact, if not mitigated and controlled at source, risks in general impact greater on the societal and institutional capacity to respond to EHS challenges notwithstanding the ethical need for global stewardship. As such a quantitative value in the form of an Environmental Sustainability Index (ESI) was created several years ago. The index is perhaps more subjective although it represents itself as an objective figure – its value lies in the awareness (especially to executives who frequently work with numbers) that it brings about some specific rationalizations of a globally complex issue to digest, and at the very least it can help, if used effectively, to get leadership in organizations to make better informed/aware objective judgments.

These sentiments are also shared by Hart (2007),[14] who talks of the new "sustainable global economy". He proposes that organizational leadership may consider three stages of implementing a green strategy, starting with pollution prevention, followed by product stewardship and looking at product life cycle impact and then the investment in cleaner/environmentally sustainable technologies. This commands a longer-term viewpoint on risks and opportunities, especially for organizations involved in manufacturing and production. Moreover, unless this is integrated into the organization's strategy, which is ultimately driven by the BoD, it is unlikely to be done effectively.

This is consistent with the growing notion of the shift from "traditional industrialism" to "natural capitalism" as described by Lovins et al. (1999).[15] A real financial value in optimization of resources with available technology improvements and the rising price of both raw materials and waste management/disposal means that environmental stewardship goes beyond doing the right thing – it makes business sense.

As discussed in a thought-provoking publication,[16] CSR, EHS, sustainability and SA issues have led to the emergence of a new profession, "the chief sustainability officer". Significantly high level issues that organizations have to address and the pressures for change are driven by EHS, sustainability and the regulations which put greater vicarious liability on the organization. On the other hand there are good incentives to change, which include enhanced brand image/reputation; decreased costs associated to insurance, losses and fines; a greater protection of assets; and increased efficiency of both the plant and its people.

In conclusion, it would seem that matters that relate to the organization behaving like a good citizen, showing that it is socially accountable and adheres to its corporate social responsibility in general, have a positive impact on EHS, which is part of protecting people and the environment. This is true in developed and also developing economies. With greater globalization of business in general around the world, it seems that the standards, or better still the expectations for corporate social responsibility, are also becoming very similar.

As such, directors must expect that the executive management team should present their strategy for aspects that relate to CSR, EHS, SA and sustainability. Today, with the Global Reporting Initiative (GRI), it is easier to establish industry benchmarks with performance and calculate meaningful and materially significant sustainability indices for industry segments and companies.

Organizations that are reporting against the GRI 4 standard today or another standard must have their board of directors involved to some extent in understanding the materiality assessments undertaken, the kind of indicators selected, the targets set and the performance of the organization overall within the sustainability frameworks. In many listed companies today, the sustainability index is considered by investors along with other investor and financial indices and ratios.

References

1 Bibbings, Roger (November 2008): "When the Going Gets Tough . . .", *Parting Shots*. Royal Society of Prevention of Accidents (RoSPA), Pages 50–51.
2 O'Connor, Tray and Young, Jackie (February 2008): "Sustainable HSE Solutions – A Saudi Arabian Case Study", ASSE-MEC-2008-05, American Society of Safety Engineers – Middle East Chapter Conference and Exhibition, Bahrain, Page 32.
3 Al Hajri, Mohammed (February 2012): "Extending Safety Off-the-Job Pays $'s", ASSE-MEC-2012-57, American Society of Safety Engineers – Middle East Chapter Conference and Exhibition, Bahrain, Page 405.
4 American Society of Safety Engineers (ASSE) (2010): "Safety and Health Sustainability Index Taskforce", Working Paper – Edited on 26 May 2010.
5 Asbury, Stephen and Ball, Richard (2009): *Do the Right Thing – The Practical, Jargon-Free Guide to Corporate Social Responsibility*. Institute of Occupational Safety and Health (IOSH), Leicestershire, UK.
6 Health & Safety Executive (HSE) (2006): "Defining Best Practice in Corporate Occupational Health and Safety Governance", A Report Prepared by Acona Ltd for the Health and Safety Executive Research Report 506.
7 Kotler, Philippe and Lee, Nancy (2005): *Corporate Social Responsibility – Doing the Most Good for Your Company and Your Cause – Best Practices From HP, Ben and Jerry's and Other Leading Companies*. John Wiley and Sons, Hoboken, NJ.

8   Schwartz, Peter and Gibb, Blair (1999): *When Good Companies Do Bad Things – Responsibility and Risk in an Age of Globalisation.* John Wiley and Sons, Hoboken, NJ.

9   Maclagan, Patrik (1998): *Management and Morality.* Sage Publications, London.

10 Bain, Neville and Band, David (1996): *Winning Ways Through Corporate Governance.* Macmillan Business, London.

11 Social Accountability International website, www.sa-intl.org/

12 Epstein, Marc J. (2008): *Making Sustainability Work – Best Practices in Managing and Measuring Corporate Social, Environmental and Economic Impacts.* Foreword by John Elkington and Herman B. Dutch. Leon and Green Hay Publishing, Berrett-Koehler Publishers Inc., San Francisco.

13 Bell, Simon and Morse, Stephen (2008): *Sustainability Indicators – Measuring the Immeasurable*, Second Edition. Earthscan Publishers. First published in 1999.

14 Hart, Stuart L. (2008): *Beyond Greening: Strategies for a Sustainable World.* First published in Jan–Feb 1997, Reprint 97105, Harvard Business Review on Green Business Strategy, 2007, Harvard Business School Publishing Corporation, Harvard Business School Press, Cambridge, MA.

15 Lovins, Amory B., Lovins, Hunter L. and Hawken, Paul (2007): *A Road Map for Natural Capitalism.* First published in May–Jun 1999, Reprint 99309, Harvard Business Review on Green Business Strategy, 2007, Harvard Business School Publishing Corporation, Harvard Business School Press, Cambridge, MA.

16 Luikenaar, Anneke and Spinley, Karen (2007): "The Emergence of the Chief Sustainability Officer – From Compliance Manager to Business Partner", Heidrick and Struggles.

8 The effective board and its relationship with the CEO/MD of an organization

To understand more how effectiveness of EHS leadership is defined, the need to understand where EHS actually resides in policy, principles and mandates becomes critical. It is equally important to understand not only how the shareholders (and stakeholders) define these standards but also how they ensure that such individuals are prepared to take on such responsibilities. What are the competencies required for this? What is the knowledge they should have? This is especially important in the context of high risk/high reliability organizations.

Whilst we see development of standards for boards, perhaps what may have been a more ceremonial, prestigious and less functional role of the BoD in the not-so-distant past is fast changing. These changes included the structuring of the board, the processes and behaviour; managing the board; the board role; understanding the board's role in corporate strategy; and finally the corporate board and the law.[1]

Some research indicates that the concept of "independence" of the board director and its inevitable value to the corporation may be contested. The research indicates that the less independent but more experienced, knowledgeable and connected directors would be of greater value to the CEO/MD and the organization.[2] The roles have changed over the last 20 years from being directors serving as a co-optive mechanism to access vital resources, as boundary spanners and as enhancing organizational legitimacy more towards linking the organization to the external environment, coordinating the interests of stakeholders, controlling the behaviour of management to ensure the organization achieves its objectives, formulating a coherent strategy and so on.

The first step of reform in corporate governance was to separate the role of the CEO and board chairman (where the board chairman is an independent director). This is especially pertaining to listed companies. Independence is critical to the board's objectivity, especially in these three key areas, as suggested by Millstein and MacAvoy (2003):[3]

1   To identify the issues it should focus on and the strategic issues of importance;
2   To obtain information that it needs to assess management performance – with respect to the chosen strategy, which includes adherence to codes of conduct;

3   To ensure that the management's efforts, as they put it, do not "obfuscate" important issues or information needed and thereby hinder the board's ability to fulfil its responsibilities and be as effective as it should.

Furthermore, in their "Intellectual Capital Model of the Board" model, they define the four key board roles to include:

1   Monitor and control;
2   Access to resources;
3   Strategizing; and
4   Advice and counsel.[2]

This is consistent with Al Hashmi (2014),[4] whose pilot study work determined that most managers saw the top four roles of the BoD to be:

1   Vision, mission and strategy setting;
2   Governance and oversight (financial and non-financial);
3   Monitoring company performance – internal controls;
4   Business continuity planning and monitoring.

The differences seem to be a more supportive role in the Western context as opposed to a more governance and oversight role in the Middle Eastern context.

There is a central and pivotal role for the chairman. The chairman's role has changed to become one subject to greater scrutiny from stakeholders, shifting from a prestigious, yet gentle way to complete a successful business career to a more involved, engaged, empowering and highly critical leadership role. Chairmen need to provide strategic counsel to the CEO and encourage board members to engage in productive critical discussions.[5] They need to develop that dynamic partnership between the CEO (and the executive management team) and the board. They need to develop a strong talent bench, ensuring that the senior executive team are working effectively with the CEO and enabling them and, in time, also mentoring them or ensuring that the right environment exists to develop the executives (and non-executives) of the future. **They need to exercise authority with empathy, recognizing that governance is as much about people as it is about process.**

Finally, one very important skill of the chairman is to command respect to ensure an effective balance of collective board strength and prevailing executive culture. Naturally, executives will concentrate on areas of opportunities to improve performance in their own areas and careers. The chairman has to be a person who is seen as non-partisan, fair and just in his/her judgement and a personality who derives the respect of the board from his/her ability, thoughtful considerations and dexterity in resolving complex issues for the company. Thus the chairman needs to ensure that the board maintains an effective long-term analysis view on such actions and not get so close to the operational details that it obscures their objectivity.

If some view the CEO as the "chief risk officer", then the chairman can be considered the "chief risk governor" for the organization. More recent thinking, as suggested by some, is that the CEO is to work with the board, regardless of structure, as a strategic partner.[6] Others explain that when reviewing literature on the strategic role of boards, at times it appears that inconsistent resource allocation creates the role conflicts between board members and executive leadership, **and this can have the significant impact of an impaired view/ judgement on strategy implementation.**[7] This is because there is no validation of action against strategy, and this can lead to a greater issue of a longer-term value proposition of the organization. This is particularly significant when resources in terms of manpower and finances are required for improvements that relate to EHS, which can be, in many cases, business continuity/risk management related and require the commitment of the leadership teams both at the board and executive levels. It then becomes imperative that a culture exists wherein the chairman, CEO, board and the executive team take a cohesive approach in recognizing, dealing with and mitigating risks faced by the company rather than protecting their limited turfs.

Nine different demographic factors were identified in one study that affects role pursuit. They include a distinctly personal and idiosyncratic approach towards chairmanship.[8] Interestingly, here chairmen focus on board dynamics and affairs and leave the enterprise matters to the CEO. Another factor is accountability spread, in which the differences between UK and US/Australian companies are mainly accountable for board performance vis-à-vis accountability for board and company performance. In the study, persons interviewed explained that the role of the lead independent director (LID) or senior independent director (SID) was relatively too new to the board concepts in the USA and UK, respectively. Many agreed that their existence did help balance board dynamics, especially those relating to the tension on certain issues between the board and the executive team due to their independence. This is important in the context of critical risk management issue-related decisions, which can become quite subjective and require a well-facilitated debate.

The study concluded that the **CEO/chairman relationship remained the single most important determining factor that ultimately impacts the performance of the board and company**.

In terms of board effectiveness there are four types/levels[9] – the first type is the "basic board", which satisfies the minimum requirements for governance and compliance. They ensure the implementation of key board processes. The second type is the "developed board", which goes beyond governance and compliance and develops a more forward-looking philosophy which develops the members' competencies and capabilities and ensures alignment with company strategic objectives. The third type is the "advanced board", which additionally looks at high performance and has members with not only a forward-thinking outlook but a better global mindset and operates within the global networks. These boards have generally higher levels of emotional intelligence, greater organizational strategic engagement and ERM.

There is a significant step change from the second to third type, as the behavioural leadership development and diversity of exposure of individuals is required. In the fourth type, a "world class board" encapsulates the traits of governance, compliance, forward looking and high performance, except that they also have a board with a breadth of insights, depth of knowledge, diversity of ideas and strength of processes, and ultimately they create greater sustainable shareholder value. This board is very rare, and both at an individual and collective level are able to add great synergetic value to the CEO, the executive team and the whole organization – especially with their insightfulness and continual improvement.

Ultimately, "A Board should possess enough collective knowledge and experience to promote a Board perspective, open dialogue, and useful insights regarding risk" (Deloitte 2011, page 4).[10]

Board directorship research suggests that the average number of members in a typical British board is 8; in France it is 14; in Germany it is 19, and the average membership in the GCC has shifted from an average of 8 in 2009 to 9 in 2011. Note that the UK and European boards are of listed organizations. In the GCC the shift from 2009–2011 has gone from 46% to about 65% respectively. In an extensive benchmarking board study undertaken on the oil and gas sector by PWC (2010), they noted that between international oil companies (IOC) and national oil companies (NOC) the boards varied in size, ranging from 9 to 15, and the range of finance specialists on boards ranged from 15% to 40%, whereas industry experts ranged from 10% to 50%, with the higher number of industry experts in the NOCs, which is interesting. IOCs had in general a greater number of independent directors appointed as opposed to institutional or government-appointed directors.[11]

An important balance is required, as whilst the independent (i.e. non-executive director and one who has not been appointed by a partner in a joint venture or private joint stock company) brings in a "less-biased" view, a level of harmony is required between board directors, who should all bring different perspectives based on their experience, specialist knowledge and their general market knowledge. This is important with EHS governance matters as well as any other board debate and decision.

It is highly recommended, however, that at least two members of the board are executives, and this is generally the CEO and the CFO. This ensures a greater connection between the business operations and the strategy development and is referred to as the "mixed board".[12] It would follow that high risk/high reliability organizations should consider having boards with a sound understanding of operational risk management.

Boards have a very complex role of being simultaneously entrepreneurial and exercising prudent control; sufficiently knowledgeable about the business whilst standing back from the day-to-day workings in order to retain an objective and long-term view; sensitive to the short-term pressures whilst being informed on the longer-term implications; knowledgeable of the local issues whilst maintaining a clear understanding of the more international aspects; and focusing on the financial performance whilst acting responsibly towards all stakeholders.[13]

As such, a certain degree of care and diligence is expected from all directors, who must carry out their functions with reasonable skill, care and diligence, and they may be liable if they are negligent; and a higher standard of performance is required of a director who may possess particular skills or professional qualifications.

It is good practice for boards to have clearly stated in the Memorandum and/ or Articles of Association (MoA) and/or (AoA) the powers of the board directors and the chairman. This is important, as this is where executive and non-executive roles may overlap. CEOs should be allowed to demonstrate their leadership and management, whilst the board should be able to "interfere" should they feel this is part of the prudent corporate control. Whilst many examples may be cited, especially when it comes to financial decisions, our focus in this book is more on the operational and non-financial performance of organizations.

In defining the "new energy executive", Csorba (2010)[14] explains that executives, especially in the high risk energy industry (oil and gas etc.), must:

1   Be effective risk managers;
2   Develop integrated decision-making skills with a balanced approach to operational, financial and EHS decisions;
3   Be accountable and self-effacing, like any leader who must accept responsibility – to see leadership as a responsibility rather than a privilege. Leaders at both CEO/MD and board level must accept accountability for things when they go wrong;
4   Be authentic communicators, and this includes being honest, transparent and clear in their communication;
5   Be involved and committed for continuous people improvement, and continuous development and improvement comes with a culture developed within an organization driven by the people at all levels; and
6   Possess high levels of emotional intelligence, and this becomes important in dealing with people with sensitivity and empathy.

Communication during and after a crisis is such an important skill of spokespersons, HSE professionals and very senior managers alike. This was clearly seen in the BP Deepwater Horizon congressional hearings with the then-CEO in 2010. The ability of leadership to stand before tribunals after an incident emphasizes the point that leadership must be ready to answer questions, otherwise they will fail, and this will have implications not only on them or their organizations but surely, as was seen with the BP incident, an impact on the whole industry. A personal discussion with a senior director of a major upstream organization based out of India revealed that the delay in the commissioning of Deepwater upstream exploration was more than 16 months, when the Sri Lankan government withdrew their approval to operate until this organization was able to reassess independently and re-submit updated risk assessment studies in the wake of the Deepwater Horizon incident, and the losses this caused were in many millions of dollars.

Integrating the governance, risk and compliance functions to align with business objectives and drive efficiencies is a critical success factor of modern organizational leadership and boards.[15] More is discussed in the chapter on risk management, but it is critical to note that whilst the process of integration may seem straightforward and simple, getting the functional heads and all these departments to work with each other must be something that is driven from the board. The reasons for this are some of these committees work directly with the board, and whilst the functions report operationally to the executive management, they functionally, at times, report to the board.

References

1 Goodpaster, Kenneth E. and Matthews, John B. (January 1982): "Can a Corporation Have a Conscience?" *Harvard Business Review*.
2 Nicholson, Gavin J. and Kiel, Geoffrey C. (2004): "Breakthrough Board Performance: How to Harness Your Board's Intellectual Capital", *Corporate Governance*, 4(1), 5–23.
3 MacAvoy, Paul W. and Millstein, Ira M. (2003): *The Recurrent Crisis in Corporate Governance*. Palgrave Macmillan, New York.
4 Al Hashmi, Waddah Ghanim (2014): "Senior Managers' Perceptions of the Roles and Responsibilities of the Board of Directors Towards Health and Safety in High-Risk, High Reliability Organizations in the Middle East – An Exploration Study", The ASSE-MEC PDCA 2014, 17th–20th March 2014, Manama, Kingdom of Bahrain – Publication Paper in the Conference Proceedings.
5 Heidrick & Struggles (2010): "Purposeful Partners", UK Board Practice Chairman Series, 2010, Heidrick & Struggles UK.
6 Favro, Ken, Karlsson, Per-Ola and Neilson, Gary L. (2011): "CEO Succession 2012 – The Four Types of CEOs", Booz & Company Annual Study of Turnover, Features, Special Report, Strategy + Business, Issue 63, Booz & Company.
7 Brauer, Matthias and Schmidt, Sascha L. (2008): "Defining the Strategic Role of Boards and Measuring Boards' Effectiveness in Strategy Implementation", *Corporate Governance*, 8(5), 649–660.
8 Kakabadse, Nada K. and Kakabadse, Andrew P. (2007): "Chairman of the Board: Demographics Effects on Role Pursuit", *Journal of Management Development*, 26(2), 169–192.
9 Gwin, Bonnie W. and Vavrek, Carolyn (2011): "Three Critical Questions You Should Ask About Your Board Evaluation", Directors and Boards, Heidrick & Struggles Governance Letter, 4th Quarter.
10 Deloitte (2008): "Perspectives on ERM and the Risk Intelligent Enterprise", The Enterprise Risk Management Benchmark Survey, Deloitte Development LLC.
11 Heidrick & Struggles (2009): "Boards in Turbulent Times", Corporate Governance Report 2009, Heidrick & Struggles UK.
12 Heidrick & Struggles (2011): "Challenging Board Performance", South Africa – Supplement to the European Report, Corporate Governance Report, Heidrick & Struggles South Africa.
13 Institution of Directors (IOD) (1999): *Standards for the Board – Improving the Effectiveness of Your Board (Good Practice for Directors)*, Edited by Tony Renton. IoD and Kogan Page, London.
14 Csorba, Les (2010): "The New Energy Executive – What the BP Oil Spill Reveals About the Essential and Defining Attributes of the Next Generation of Leadership", Heidrick and Struggles. First published in the *Houston Chronicle*, July 18, 2010.
15 PricewaterhouseCoopers (PWC) (2005): "Corporate Governance", Governance, Risk and Compliance Series – Connected Thinking, Global Best Practices, Pricewaterhouse Coopers Publications.

9 Organizational structure and effective safety communication

In this short chapter, the author explores organizational structures and how they can have a direct bearing on how safety is managed and led within an organization. Communication is a crucial aspect of informing, reporting, monitoring and eventually engaging leadership in supporting initiatives and change as and when needed to control risks when they arise.

Safety must be managed throughout the organization and led from the top, where strategic risk reviews must be undertaken of all the operations, involving the leadership team and preferably involving the board directors.[1] Defining the key EHS responsibilities and accountabilities within the corporation is critical. The role of the EHS managers/advisors must be defined. They would be responsible in the context of high risk/high reliability organizations to help manage risks in such a way that they bring sustainability to the business.

The impact of major incidents due to lack of any safety controls, for example intoxicated workers on the site of plants in the late 1800s and early 1900s, drove DuPont to develop better safety regulations. Enhanced safety provision has arisen from improvements in the past 50 years to both physical (technology-related) and procedural (administrative) methods. This was further enhanced throughout the industry through effective sharing of incident investigation findings within the industry.[2]

However, it is reported by some that many of the employees surviving the blow-out incident in the Deepwater Horizon incident with BP indicated that management routinely dismissed warnings and documented procedures to hasten making the well productive. So it was an unsafe company culture which developed.[3] It thus becomes necessary for the board members to have a process of effective risk recognition and take adequate measures by defining an organizational framework that ultimately mitigates these risks. EHS data is very important to collect and analyze in organizations, and organizations must use this information to make improvements.[4] They collect both passive and active date/indicators, commonly called lagging and leading key performance indicator (KPI) related activities, respectively. Industry must be able to effectively measure, monitor and control. That is why KPI reporting is one of the best EHS communication methods and helps senior leadership make decisions.[5] EHS KPI summarized data reports with trends go to the BoD of all major high risk/high reliability organizations.

Good EHS performance is attributed to a particular practice or initiative from the management team.[2] The balance between leading and lagging indicators is important, building the KPI system on the basis of the Global Reporting Initiative (GRI) or the Balanced Score Card (BSC) integrated management systems and making it part of the business management systems and intelligence rather than a standalone system to measure EHS effectiveness.[6]

In the Eastern Petrochemical Company (located in Saudi Arabia), EHS performance is a key performance measure of the organizational performance. Senior managers are held accountable for safety performance, as any incident is caused by a lack of effective risk management and control.[7]

Moore (2008)[8] notes that using the Centre for Chemical Process Safety (CCPS) definitions explains the differences between the *lagging metrics* – retrospective metrics based on incidents that meet a threshold of severity that should be reported – and *leading metrics* – a forward-looking set of metrics which indicates performance of the key work processes, operating discipline and layers of protection that prevent incidents. In the Baker Panel recommendations, an improvement in the PSM in all industries with benchmarking was a fundamental improvement requirement.[9]

Reporting EHS performance can be a contentious issue for managers. The importance of the reporting lines and structures within organizations when it comes to EHS managers is very much linked to the empowerment they have to manage risks and openly and effectively prevent loss. However, even if empowered, EHS managers must also try to acquire the right business acumen in order to better present data and information in such a way that management, be it operational or executive, better relate EHS to the business itself. Whitaker (2007), in some very interesting and relatively unique research, tackles the issue of reporting of EHS within an organization and discusses the advantages and disadvantages of centralization, decentralization and hybridization as well as matrix organizations.[10] An emphasis is on which models work best depending on the organizational structure, size and locations of operating sites. The importance of EHS reporting to the highest authority within an organization to ensure the right information is sent up in a timely manner is reconfirmed.

Practitioners must report to the organizational leaders. This is not only so that they may help in directly implementing their commitments; it also, as this reporting line in itself demonstrates to everyone within the organization that this position is a senior one, shows that the position carries equal importance with all the other operational, technical and financial functions.[11]

Risk management governance frameworks for the national O&G industry should be a management process, whilst HSE is a business support function requiring the involvement of committees to help direct efforts effectively within an organization.[12]

Finally, on the matter of effectively communicating the safety message, the EHS practitioner has a key role in this, and practitioners must be able to do this with visual and organizational support.[13] Therefore, leaders in organizations must be seen and heard supporting EHS. Moreover, the BoD must monitor the performance of an organization in every way. This includes EHS performance, as the impact on sustainability of the business, reputation, assets and most importantly people can be very high.

References

1 MacLean, Richard and Row, Rick (March 2011): "Resource Wars – What's Your Battle Plan", Air and Waste Association, Pittsburgh.

2 Freibott, Bernd (February 2012): "Sustainable Safety Management: Incident Management as a Cornerstone for a Successful Safety Culture", ASSE-MEC-2012-49, American Society of Safety Engineers – Middle East Chapter Conference and Exhibition, Bahrain, Page 338.

3 Sullivan, John (June 15): "A Checklist for Predicting Corporate Disasters – Is Your Firm the Next BP?" Internet Article, available from http://drjohnsullivan.com/a-checklist-for-predicting-corporate-disasters-is-your-firm-the-next-bp/

4 Katoty, Dehbashish (February 2012): "Developing a Hazards Learning Culture – Interpreting Information From the Electronic HSE Reporting System of Kuwait Oil Company", ASSE-MEC-2012-57, American Society of Safety Engineers – Middle East Chapter Conference and Exhibition, Bahrain, Page 405.

5 Travers, Ian (2012): "Key Process Safety Performance Indicators – A Short Guide for Directors and CEOs", Health and Safety Executive, based on the HSE Guidance Developing Process Safety Indicators – A Step-by-Step Guide HSG254, HSE Books, available from http://hse.gove.uk.

6 O'Connor, Troy (February 2010): "SHE Leading and Lagging Indicator Dynamics", ASSE-MEC-2010-35, American Society of Safety Engineers – Middle East Chapter Conference and Exhibition, Bahrain, Page 273 (Ref 2).

7 Al Jaffar, Ahmed Hussein (February 2010): "Is Safety Common Sense? A Systematic Approach to Safety for Project Management and Construction", ASSE-MEC-2010-33, American Society of Safety Engineers – Middle East Chapter Conference and Exhibition, Bahrain, Page 262.

8 Moore, David A. (February 2008): "Process Safety Leading and Lagging Metrics", ASSE-MEC-0208-30, American Society of Safety Engineers – Middle East Chapter Conference and Exhibition, Bahrain, Page 190.

9 Baker, James (III), Bowman, Frank L., Glen, Erwin, Gorton, Slade, Hundershot, Dennis, Levison, Nancy, Priest, Sharon, Rosentel, Tebo, Paul, Wiegmann, Douglas and Wilson, Ducan (January 2007): "The Report of the B.P. U.S. Refineries Independent Safety Review Panel", available from www.csb.gov.

10 Whitaker, Joseph Mathew (10th September 2007): "How EHS Managers Can Influence Environmental Excellence Within their Organization", Graduate Thesis submitted in partial fulfilment of the requirements for the degree of Master of Science in Environment, Health & Safety Management, Department of Civil Engineering Technology, Environmental Management & Safety, Rochester Institute of Technology, Rochester, NY.

11 MacLean, Richard (2007): "Environmental Leadership – Get Organized", *Environmental Quality Management*, 17(2), 95–98. doi:10.1002/tqem.20258.

12 Booz & Co (2010): "Oil and Gas Sector and Corporate Governance Benchmarking", Booz and Company Consultant Research Report Presentation, Version 2, Beirut, Lebanon.

13 Bibbings, Roger (July 2003): "Hearsay and Heresy", *Parting Shots*. Royal Society of Prevention of Accidents (RoSPA), Pages 32–34.

10 Exploring the themes impacting EHS governance and leadership

In this chapter we go through research that was conducted on more than 20 organizations in the Middle East[1] and in which certain themes have been explored further through many interviews. This chapter also explains the importance of understanding further the sub-themes that emerged from a more detailed interrogation of the qualitative data that was collected. These areas are what are seen to be the major influencers on EHS leadership/governance matters in general.

Background

Thirty senior leaders ranging from CEO to MD, board director and executive director, representing 20 different high risk industries, were interviewed through a semi-structured discussion. The process was extremely insightful as the results give greater depth of meaning in terms of personal perspectives from leaders about the notions of EHS leadership and governance. Some of these interviews lasted for more than one hour, and through the transcription and a process of structured content analysis the following sections highlight the key areas – and their meaning for many of the senior leadership – which shape the policies and implementation of effective EHS leadership and governance in their organizations.

Developing safety culture and communications

Leaders were asked about the areas of the development for an effective safety and EHS culture in their organization and the significance of communications within their organizations. In contrast to the literature, an EHS culture is seen by many leaders as an organizational matter, and the interviews showed many statements that supported notions of (1) the importance of developing a culture of rewarding safety excellence and good EHS practices; (2) instilling a no-blame culture which promotes transparency, especially in reporting incidents; (3) how a culture can be developed by making EHS the first item of discussion in any meeting; (4) developing organizationally integrated behavioural-based safety programmes and so on.

The importance of understanding some of what we refer here to as the sub-themes in all the analysis undertaken brings about both a greater understanding and appreciation of the matters that influence leaders in demonstrating actions within this theme.

However, it would seem that whilst many of the senior leaders spoke about having a culture of EHS within their organizations, it is clear – except for a very few leaders, more from internationally based organizations – that this culture was being driven at their executive level rather than the BoD.

In saying this, many leaders did say that the BoD were generally supportive of efforts that the executive management exerted towards developing a culture of safety and EHS. They set the overall direction for the organization either directly (or explicitly through expectation setting) or indirectly (more implicitly through seeing EHS matters as an integral part of performance). One matter which evolved from the interviews was the clear influence of international best-practice standards, especially in larger organizations which may be based regionally but operate in a more international environment.

A total of seven sub-themes evolved from investigating these important themes. They include:

- Leadership creates a culture and influences/sets the agenda for change. Many leaders said that was important at the board level but more critical at the executive management level.
- Reporting, transparency and "no-blame" culture. Many leaders said that developing a very open and trusting organizational culture was never easy. They stressed that this is even more challenging in larger and diverse organizations. Developing this needed more intervention at the supervisory level of the organization to motivate employees to be more open and transparent, and this at times is difficult, especially with performance management systems that placed greater emphasis on the importance of positive lagging EHS indicators being reported.
- Changing the behaviours is key to changing the EHS culture. Here many leaders emphasized that EHS culture must be part of the organizational operating culture, and only when this is achieved do the behaviours of employees lend themselves to a more positive culture where EHS is seen as a true and respected value.
- General EHS culture where EHS is a value. In fact, an organization which sees EHS as a value rather than a function is very important, and this is discussed further in this chapter and the next.
- Compliance and consequence management. Many leaders mentioned striking the right balance between encouraging compliance and managing non-conformances.
- Moving from high risk to high reliability organizational EHS cultures. This shift in changing a culture is what most mature organizations strive for. This is very consistent with what was discussed already in Chapter 2. Culture development through learning organizations is also something that

some leaders alluded to, but this was with leaders of larger and established organizations.

- The impact of international standards along with the size of the organization was a very important aspect that was raised and which has been discussed later while covering the model. One of the external factors of an organization is this expectation for it to raise its standards to an internationally acceptable practice level.

Safety/EHS leadership

It would seem that much of the safety/EHS leadership seems to be mainly driven by the CEO and his executive team. Senior leaders saw, in general, the role of the BoD as being more supportive rather than directive when it came to EHS matters, as they cited many examples of an operational nature. They reverted in many interviews to giving examples from the working level, which seems to indicate that they see many aspects of EHS as operational rather than strategic. It also may mean that they found it easier to articulate examples of EHS matters more in an operational rather than from a strategic context.

In saying this, some recognized the power – especially within the Arab world, which has a very traditional tribal culture – that the leadership's words and actions have, going a very long way towards influencing change. EHS culture and leadership are highly related, and in many of the statements many leaders implied that good EHS leadership was demonstrated, in an effective EHS culture, with transparency; with high levels of commitment through visible leadership from managers and workers alike; with a shared vision and belief in EHS as an important core organizational value throughout the organization; and so on.

Under this theme and upon a closer review of the statements made, three sub-themes evolved, which included:

- Demonstrating safety/EHS leadership in terms of visible actions from leaders. This is why some leaders saw this to be a natural role of the executive and not the board – visible leadership where safety/EHS was being driven from the top and where there is a tangible demonstration of leadership actions.
- Board leadership vs. executive leadership is an important and really fundamental aspect to define for many leaders. The board once again needs to be supportive and motivate executives to demonstrate good EHS leadership along with those who saw their demonstration of EHS leadership.
- Making safety/EHS a core value. Many leaders said it was many times more difficult to make EHS a core value.

What is also fundamental is that it was recognized that EHS leadership played probably the single most important role in creating a safe and reliable organization.

Influence and accountability

More than half of the statements analyzed in the interviews related to these three top themes: *EHS leadership, EHS culture and communications* and *influence and accountability*. Senior leadership expressed very strong views on this. Whilst many of them felt that ultimately they would be accountable for EHS incidents, they did explain (and with some citing specific examples) that they would be held accountable for wilful negligence. In the survey two statements were evaluated: (Q1) *Each board member needs to appreciate that his/her actions/decisions (where applicable) should reinforce the health and safety policies and statements with no contradiction* and (Q2) *The CEO/MD should reinforce directives given by the BoD even when they may not be aligned with the health and safety policy.*

There was a marked difference between the level of agreement for Q1 and Q2 by the oil and gas group, where they were in agreement with the BoD having to appreciate their actions/decisions, whereas there was less agreement with the CEO/MD accepting and reinforcing BoD directives which were not aligned with the policy.

That difference was not seen with the responses of the non-O&G group. This may be better explained with the greater level of autonomy, empowerment and accountability that senior leadership feel in the oil and gas sector, which is perhaps more self-regulated than other non-oil and gas industries such as shipping and aviation.

Three sub-themes evolved under these themes:

- CEO/MD accountability cannot be delegated, and this was something that many leaders recognized and were greatly concerned with. The delegation of responsibility needed to therefore be undertaken with due care. This they highlighted as the importance of employing competent and responsible managers in their organization.
- Stakeholder influence on accountability: this is in terms of the expectations that shareholders and other stakeholders have from leaders in organizations, be they executives or not. It was felt that this also shaped and defined the accountabilities.
- Board accountability was difficult to clearly define, as it is less straightforward. Clearly the executives felt very strongly that the board of directors (BoD) needs to accept both formally and publicly their collective role in providing health and safety leadership in their organization.

Generally, a good discussion emerged in many of the interviews, which identified the great challenge of where accountabilities lie between the more "collective" board and the more "individual" CEO/MD. It is this grey area which makes this research so significantly important. In the making of high reliability organizations, the clear partition of board and executive leadership actions is at the very foundation of this EHS leadership and governance debate.

This is consistent with both the literature (Roy 2010;[2] Qadir 2010;[3] Ahmed 2008;[4] Anderson 2008;[5] and Roger et al. 2009[6]) and the statements coming from the interviews, where the research participants all indicated that account-ability and responsibility was with the senior leadership including the BoD.

However, this area remains somewhat grey, as many senior leaders also saw ultimate responsibility/accountability with the CEO/MD. The differences in views are explained by three aspects:

1   BoD is responsible and accountable for governance, as opposed to CEO/MDs, who are responsible for action;
2   There are differences in opinion with respect to the board's collective responsibility rather than individual responsibility (perhaps once again influenced by a more Eastern collective culture rather than a more indi-vidualistic Western one); and
3   Many CEO/MDs expressed their discomfort with BoD members involv-ing themselves with some of the "day-to-day" decisions; therefore, those who saw EHS more as an operational matter supported a clear distinction in the roles, especially when they also indicated that the BoD members did not understand enough about the business.

Ultimately, one of the emergent aspects from the interviews, especially with respondents who supported a detached board structure (which will be discussed later on), is the separation of roles and delineation of the roles and responsibilities, insisting more on the accountability of the CEO/MD and limiting the account-ability of the BoD to being effective in their "governance and oversight".

Moreover, in the emerging model the relationship is explained by the fact that morality, which is a personal factor, interacts with legal imperatives for EHS, which is a more socio-economic (external) factor on the organization. This is discussed further in Chapter 13.

Monitoring EHS performance

It would seem, with respect to the BoD, almost all senior leadership agreed strongly on the core role of oversight through monitoring EHS performance.

Upon further interrogation of the qualitative data, three sub-themes emerged:

*   Effectiveness of monitoring EHS performance;
*   Frequency of monitoring EHS performance; and
*   Expectation/standardization from the BoD on reports.

From the interviews conducted, more than 10% of the statements analyzed for themes related to monitoring. Boards are seen to be getting more and more involved today with growth in awareness of the importance and impact of EHS. Many senior leaders even saw this role as limited due to the lack of knowledge and competence of directors in certain organizations. This clearly impacted on the "effectiveness" of monitoring. Many respondents also explained that whilst boards monitored EHS performance, their doing so added limited value.

The general variation in reporting trends (frequency of reporting) within these industries and, in fact, within the same industries in the GCC vary from reports being issued weekly for board meetings held on a monthly basis and reports being issued monthly for board meetings taking place two to four times a year. The reporting is generally part of the practice of all senior leaders and their boards. This is seemingly an established culture, and many leaders said that this was a standard practice in the industry to give the board some assurances through (mostly standardized) effective periodic reporting.

In the discussion with leaders from the oil and gas sector, some commented on the importance of going back to process safety management-related indicators rather than personnel safety indicators. They have recognized that the focus on lost-time incidents and other such lagging indicators added less value to helping boards appreciate the level of effort and resources that are required to sustain high performance and lowered incidence rates. Some said that proactive key performance measures and what is termed "leading" indicators were much more important to link effort with higher EHS performance.

It is to be appreciated here that boards in most organizations get involved in EHS when something goes wrong, and they want to understand what went wrong and what is being done to both deal with the consequences and prevent reoccurrence when they are asked to approve large budgets in order to spend to improve EHS or otherwise or when presenting performance data for the last reporting period. When there is focus on reactive KPIs rather than proactive KPIs, the perception of EHS becomes somewhat of a negative issue focus area.

This must change in the future because if the board is to play a more positive active role, it has to play a more proactive role advising and supporting the executive leadership, as noted by various scholars (e.g. Kakabadse and Kakabadse 2007;[7] Nicholson and Keil 2004[8] etc.). It must also base its expectations of the executive on proactive performance in EHS and recognizing the efforts made in preventing incidents. On the other hand their role must also be focused on motivating management in their development of organizational resilience, effective sustainable growth and more effective organizational EHS cultures through EHS leadership.

Risk management

The common theme-related questions most leaders disagreed with or least agreed with is related to risk management. Questions related to leadership and board, risk management strategies and practices draw differences in opinion and outlook when compared to best practices, as in the literature reviews (e.g. Deloitte 2008;[9] O'Conner 2012;[10] Dufort 2013[11] etc.), and this also reflects the difference in the approach and practice of different industries in dealing with risk management and control. As discussed in Chapter 6, it would seem for the analysis of data that there may be currently greater understanding and acceptance of enterprise risk management models in the oil and gas sector as opposed to the other industries.

Again this may be related to the fact that the oil and gas sector is more accustomed to being, due to the development of the industry over the years, more self-regulating compared to other industries, especially the maritime and the aviation industries, which are highly regulated and controlled. As such they developed their own internal risk management systems and were more comfortable with managing risk in the context of the board and executive management through high level enterprise risk management strategies and processes.

When the respondent feedback was interrogated more closely, three sub-themes seem to emerge, which include:

- *Risk awareness* – board directors need to be more aware of what the credible risks are and how those risks are being mitigated;
- *Risk appreciation* – board directors need to see these risks as opportunities in a business context and also appreciate that risk controls are a means for opportunity for development, value addition and growth for the organization; and
- *Risk Tolerance* – board directors have to set what is commonly referred to as "an appetite" for risk. This is based on the fact that business value is created by taking a certain risk and deciding what level of risk is acceptable to the board (notwithstanding the mitigation measures and controls in place), and this is risk tolerance.

It would seem (logical) that boards are interested in risk and understanding how they are generally being managed. However, once again the impact and the normal or best industry practice in that industry is what drives the risk management models. This is, however, highly dependent on the level of both risk awareness and appreciation.

To this end it would seem that because EHS risks are technical in their nature, the engagement of BoD members in addressing or discussing those risks is not significant. However, when looking at the overall exposure of an organization, directors look to be comforted that a certain system is in place and functioning effectively to mitigate risks, be they EHS or otherwise. This is especially the case today, where major EHS incidents relating to catastrophic failures often go back to root causes clearly related to leadership and governance failure.

It would seem also, when comparing this with the current literature on enterprise risk management (ERM), that when EHS risks present themselves as credible business (and business continuity) risks, a greater board interest and thus risk appetite is overtly expressed.

EHS awareness, knowledge and competence

It is interesting to see that more than 50% of the respondents supported mixed board structures (where the CEO/executives are also board members), as will be further elaborated on in the next chapter. One of the key elements behind this was clearly that the leaders felt that this brought about greater alignment, and as will be discussed in greater length later, it is the knowledge and competence on EHS matters that executive board members bring to the table which leaders felt was important. Two sub-themes came about in further analysis of the statements made during the interviews, and these included general awareness and the induction (on-boarding) of programmes that organizations provided or at least needed to provide to board directors.

Consistent with Van der Westhuyzen (2012),[12] who spoke about monitoring, i.e. control against set performance standards and expectations, it would seem that a good number of leaders see the onus on their organization to prepare the board directors to understand their business. There is no doubt that knowledge and competence is required in both setting the performance standards and monitoring effectiveness.

This must be a serious area of concern as, on the whole and regardless of which industry was being addressed, the level of EHS knowledge and competence with the directors was limited to the extent that, in the minds of the leaders, they added limited value to the overall effectiveness of (EHS) risk management, monitoring and advising on effective controls. However, the industries are taking effective steps to address this by arranging for induction programmes for (at least) new directors which include detailed operational overviews and EHS in various organizations. One oil refining company went to the extent of running a one-day workshop which had a significant component about EHS.

Finally, it is important to appreciate that leaders recognize the need for knowledge and competence and − even if the chairmen of boards do not directly ask for it − consistent with what has already been discussed, some of the senior leadership are proactively arranging for induction and awareness programmes. And to this end, from 15–19% of respondents saw that board structure made little difference to EHS leadership; they supported this by saying that it was more dependent on the diversity and makeup of the board. Some of the leaders interviewed basically said that if you had directors who had some industry knowledge and an operational/EHS background, they would be able to contribute more effectively to EHS matters.

Operational excellence and management systems

In a similar way to the responses on risk management, the variation in responses apparent may be explained by the fact that this is viewed as more of an operational-level matter rather than a strategic matter for the board.

Moreover, this is apparent in the significant difference when comparing the O&G and non-O&G industries. Once more this can be attributed to the fact that oil and gas is more self-regulating and that the development of mature EHS management systems aspiring to excellence was a result of broader and more general and less prescriptive laws and regulations (as compared to maritime (and transportation), aviation and manufacturing or even the construction industries).

However, as discussed earlier the development of management systems is critical to the development of high reliability organizations. A very detailed explanation was given by one senior leader interviewed as to how his organization developed their operational excellence model over time and how it was built on a set of beliefs and principles. Interestingly, in some organizations many of the EHS management systems are built on management system elements. They refer to them in different ways or call them different names. For example, VOPAK, the Dutch-based oil and chemicals storage company, calls them "Fundamentals"; Chevron, the oil major, calls them "Tenets"; Petroleum Development Oman (PDO), the Sultanate of Oman's national oil company, calls them their "Golden Rules"; and the Dubai-based diversified oil and gas company, the Emirates National Oil Company (ENOC), refers to them as their "Principles", to name a few examples.

Three areas influence the improvement in systems, and as such the board only drive this through setting their expectations from the CEO/MD and his team:

- Operational excellence – is a thinking going beyond simple EHS management systems to a system that aspires towards integrating operational resilience; business and quality excellence; sustainability and environmental stewardship; asset integrity and so on. This is generally a great effort and requires the organization's leadership to commit and drive this implementation of what has come to be known as operational excellence;
- Beyond (legal) compliance – legal compliance or similar compliance creates cultures within organizations which are described as "calculative". They lack proactivity and also to a great extent fail to create a continual improvement drive within an organization. CEO/MDs should be motivated to therefore go beyond compliance and look at the excellence that can and in fact must be achieved in their organization. Whilst organizations must "work within the regulatory framework", they should not lose focus on continual improvement in the context of creating operational excellence systems; and
- Best practice – CEO/MDs must be motivated to look at their organization's practices and benchmark them with others, and then, by looking at their own practices, to see how they can achieve better practices or best industry practices.

Operational excellence was explained by means of an integrated management system which looks at asset integrity, operational discipline and high standards of embedded EHS processes and procedures. Here the policy drives much of the management system elements embedded very strongly, as discussed in the earlier chapters, fundamentally by leadership.

Systems thinking was related to organizations believing in a fully disciplined and structured approach towards implementing management systems that obviously is due to the risks involved in EHS. It is interesting to note that this systems thinking leads to a "beyond compliance" paradigm. It is based on setting an internal set of standards and working towards meeting them for the purpose of safe and reliable operations rather than due to legal regulations and requirements.

Moreover, generally with almost all of the senior leaders interviewed, the foundation of their EHS assurance came in the form of a management system of some sort. It is critical to note four very important aspects which came about from the interviews, in the sense they would most probably agree on:

1   Having processes and procedures in place was fundamental to having an effective EHS performance with zero or fewer incidents;
2   Compliance to the rules, laws and regulations sets the minimum benchmark which you cannot go below. But your industry standards must be higher because they have the latest information drawing on the many lessons from industry;
3   Setting rules and standards within an organization creates the operating boundaries for everyone to work within and, just like a state sets rules and regulations to maintain law and order, an organization sets EHS policies and procedures to maintain operability and stay resilient and sustainable, especially if it is a high risk industry; and
4   Larger organizations (holding corporations) have developed guidelines and codes of practice for all their operating units (affiliates) to bring about a good level of compliance and operating discipline with all of their operations.

But what was also emphasized is that excellent systems required a firm and fair consequence management process in which if people did not comply they were counselled, warned and fired. It was emphasized that, without this, systems could not achieve excellence.

One important aspect in the development of management systems which we see is that the policies, procedures and guidance have become more dynamic in that they continually are upgraded and change to deal with new and emerging risks. Best practice, defined by industry benchmarks and practices, drives much of the journey towards excellence. Excellence by definition means that compared to industry norms, organizations were not only meeting higher standards, but they also had systems in place that created continual improvement. Road safety is a case in point and was discussed by more than one senior leader; this is especially an issue in the oil and gas upstream sector, where leaders explained how they have exerted a great deal of effort as an organization to study the underlying factors and, through engineering, administrative, policy and procedural controls, to eliminate the risks. In the GCC for example, as various leaders discussed, poor road safety is the No. 1 cause of fatalities in that industry.

Legal imperatives for safety

Regardless of which industry one belongs to, senior leaders recognize the significance of the legal imperatives for safety and EHS. As discussed earlier there is an interesting relationship between legal imperatives, operational excellence management systems and risk management. Clearly they are strongly linked, as the management systems would be developed on the basis of the minimum legal requirements and as an assurance mechanism for the leaders that the procedures are in line with the legal requirements within a jurisdiction.

In saying this leaders explained that they ensured that their organization applied whatever systems helped in making their operations effective, efficient and safe, ensuring only that they did not go below the legal requirements. They said that legal aspects become extremely important after an accident or incident because in any major incident in which people lose their lives or are injured, the environment is damaged and/or assets are damaged, it has to be appreciated that it is as much an operational and financial and perhaps reputational loss as it is an EHS or safety omission. Legal compliance is important in keeping a business healthy and running without losses. Two sub-themes emerged in this analysis which included:

- Core business conduct imperative and
- Basic compliance.

Legal compliance was related in the interviews to business conduct. The leader's moral duty to protect company employees was something that was a given from the responses of many leaders. But at a more basic level, compliance is critical to business from a "license to operate" perspective.

Many of the legal debates suggest that there are two dichotomous approaches[13] to the management of legal non-compliance, especially in occupational health and safety. The two schools of thought are either (1) self-regulating for compliance or (2) policing. It is not clear, from the discussions with senior leaders, if they saw this as just compliance and did not discuss such matters; if so, then it comes down to the law. Perhaps this reflects the level of involvement of major industries, even if at a regional as opposed to a global level, in the development of regulations and standards. As such the leaders did not see themselves and their organizations as stakeholders in these regulations as much as they saw themselves as just having to ensure they complied. Industry generally in the GCC, for example, is not always consulted about laws, rules and regulations (by the regulator) as they may be, for example, in continental Europe. This is, however, starting to change.

With this in mind, legal imperatives for EHS remain a compliance issue related to the organization's license to operate. It is a bare minimum rather than an aspect which drives best-in-class EHS leadership and organizational EHS cultures.

Reporting structures and hierarchies

Many organizational leaders see this aspect of reporting structures as being a policy of direct reporting and indirect reporting with respect to EHS. Clearly, when asked the question on the reporting line of EHS managers/directors with respect to the senior leadership, there were three types of generally varied responses:

1   They must report to the CEO/MD directly – as some of them put it: "*I am ultimately responsible for EHS*";
2   They can report to a member of the CEO/MD's senior technical or operational executive management team, as they said they have a system and there is good transparency. But here, ultimately, EHS/safety was seen as pertaining to technical functions like maintenance or inspections functions etc.; and
3   They really have to report to the CEO/MD because this sends a message to the whole company and the other executives that safety is at least on a par with other aspects of company operations.

Whilst reporting lines are an important aspect to understand and appreciate, it can be said that the impact on EHS leadership as such is not as significant as with the other themes. But this might be dependent on the maturity of the organization or otherwise the current structure of an organization. Thus, if an organization has a mature operating system and the roles and responsibilities for EHS are clearly defined in all jobs rather than only the EHS practitioner's job, then the reporting lines may have a certain flexibility. Reporting lines are important to ensure that the correct information is received by the organization's leaders in an accurate and timely manner.

To conclude here, the issue of direct and indirect reporting of EHS depended on the governance imperative of the organization's structure when it came to EHS. The organizational leaders who seemed to see EHS as a more strategic or at least tactical matter preferred the direct reporting structures, so it would seem that they wanted to lead EHS and be directly informed, or at least place EHS at a level such that others within the organization would appreciate that it was important to them.

Exploring the applicability of the themes/ideas that emerged from the interviews conducted

From the various interviews conducted, there were various areas and themes that emerged through some of the statements that were felt to be highly significant, to the extent that they may very well be a theme in their own right. One clear drawback of much of the existing literature around these subject areas (which address EHS leadership and governance) is that much of it has been a contribution of EHS practitioners, governance specialists, legal specialists and so on, and whilst many perspectives were explored, this remained limited in the sense that perhaps the senior leadership perspectives have not had a fair hearing. Whilst there have been publications which have gone through significant consultation with senior leadership (OECD 2012;[14] IoD-HSE 2008[15] etc.), to cite one example,[16] the disappointing reality is that the surveys conducted on UK directors on the HSE/IoD 2008 code over two years showed little improvement in awareness, readership and implementation of the code. So it may be argued that whilst they were given opportunities to review and comment, many senior leaders may not have invested the time to do so. Therefore, how much of their input guided these standards remains to be investigated.

Leadership style

Some leaders said that boards will naturally have a **reactive leadership style** because directing is a **passive leadership style** in which board members will react to the information presented to them on performance. Thus all they can do is inspire and motivate the executive management team to be more proactive when addressing EHS matters. This is not so much a criticism as much as it is a reality which has evolved through the workings of boards. As discussed earlier, some boards are more active and meet more frequently, and as such it is believed that those boards probably have a more **directive and influential style.** But as noted by Clarke (2012),[17] effective safety leadership requires both transactional and transformational leadership styles to be effective.

Role definition – segregation

The **separation of the roles of the board and the executive** was argued by some leaders as being the only way to achieve real governance. There should be a clear distinction between managing EHS by the executive and directing EHS within an organization by the chairman and his/her board of directors. This is consistent with Millstein and MacAvoy (2003),[18] who proposed a separation between the role of board and executive, a view shared with Nicholson and Keil (2004)[8] and others.

EHS – a value driver

The concept of **EHS being a business value driver** was discussed, and many leaders explained that EHS needed to be looked at in a different way, perhaps akin to quality management within organizations; EHS management in their minds brought about improvements in productivity, sustained growth and prevention of disruption to business. In high risk organizations it is also related to reliability, which is a valuable trait for an organization and provides stakeholders with a feeling of resilience. As such some leaders therefore felt that embedding **EHS as a key organizational value driver** was required by the leadership for an organization's **EHS culture** to grow and become a part of the fabric of the organization, its goals, its objectives and its beliefs.

In a very similar way, **EHS as a sustainable value driver** was also discussed, and some leaders said that sustainability gives organizations a long-term sustained value over time, which is one of the long-term objectives of a board. In a recent research presentation in India for the oil and gas industry, research showed that "Governance sets a long-term destination whereas leadership sets the road map for the set coming period under direction of BoD" (Al Hashmi 2013,[4],[19] slide 14). The sustained existence, operation and growth are all very important aspects of EHS assurance. When talking of the banking industry's corporate governance standards:

> In a world in which financial markets reward short-term reported profits, it is the responsibility of the bank's board to take care of long-term value creation, even if that means hurting reported revenue and the share price in the short term. Executives should drive the business within regulations in accordance with the strategy and manner (ethics/culture) set and supervised by the board.
>
> (Dermine 2011, page 9)[20]

Alignment and competing goals

On matters of EHS leadership effectiveness, many senior leaders explained that it is critical for this to be aligned to the vision between the executive management and the board on all EHS matters. This would naturally also deal with another important aspect or theme which emerged, which is when **EHS has competing goals with commercial/financial drivers**. It has to be appreciated that EHS will continue to struggle against financial goals in organizations, especially for investment and policy issues that may be seen by operational and marketing staff as "restricting" and as an expenditure with returns very difficult to quantify in traditional financial return on investment terms. This in turn creates conflicting priorities in business. Some conflicting priorities in business can have very negative effects on EHS. An example of this was deferring maintenance works on a pipeline to save costs and the pipeline failing as a result of corrosion. Leadership needs to assess and ensure that risks are not created and EHS in an organization is never compromised. This is sometimes a challenge, as even the most accurate quantitative risk assessments can be based on quite subjective assumptions due to the nature of risks which are related to failure rates of equipment (e.g. mechanical seal failure on a pump), natural factors (such as storms and earthquakes etc.) and human factors (i.e. human error born from another varied set of factors including lack of training, fatigue, underdeveloped safety culture etc.). Thus it becomes imperative to understand the intrinsic value of EHS goals and objectives and that those goals be adopted and appreciated by the leadership weighing the commercial/financial impact and value of EHS.

Industry influence and maturity

A very clear factor which emerged from the discussions, especially within the oil and gas upstream and the aviation sectors, was that it was a fact that industry was in itself driving a change and focus. The best practices were employed as a result of lessons learnt from different incidents and accidents or as a result of more proactive technical research as well as the basic fact that all these different companies compete with one another in delivering services and products to other businesses or to the public directly. All of this has had a profound introspective focus shift towards better EHS standards. In oil and gas, for example, a few of the leaders interviewed said that the focus of the industry on process safety management in recent years has come about as a result of recent international business-crippling incidents which highlighted to management that they have been too focused on personal safety and have neglected plant and process (hardware) safety.

Therefore the emerging global and regional influences such as trends, international standards and regional practices continue to influence and motivate/drive change within organizations and their practices. This is in general terms, but very much so with EHS matters.

Effective monitoring and analysis

Monitoring EHS performance has already been discussed. However, what was highlighted in the discussion with leaders was that the **analysis of performance** was hugely important by the board rather than just monitoring. Greater engagement and understanding of the reasons underlying the EHS performance, be it good or bad, are more important than monitoring that performance. As the BoD needs to add value to the discussion on EHS with the executive team, they need to understand why the performance is the way it is rather than merely what it is.

On board matters in more general terms, the board structure that ensures objectivity and has the right level of knowledge and a healthy working relationship with the leadership teams was very important. The structure is seen by some to be less important than the engagement of board members and board dynamics. Concerning the makeup of the board, ensuring there is a good mix and that there is diversity of experiences and knowledge was supported by almost all respondents, who probably saw that as the single most important aspect of board effectiveness in dealing with EHS matters. Whilst organizational maturity governs boards, at the end of the day the board oversight role must be very strong.

Social accountability and impact

The growing impact of socialization and social value contribution of organizations and corporations is consistent with the issues identified[21] in regards to the lack of good practice and corporate social responsibility governance in the profit-making organizations as opposed to the not-for-profit organizations. Although their research focused on the medical services sector, the matters related to **social impact and accountability** have become important, and in the GCC and the Arab world in general, this has become a matter of significant socio-political importance, with a growing demand for governments to create employment opportunities. However, another important aspect highlighted was the alignment of the organization and the business and for them to take into account people's needs for both external and internal customers.

Learning organizations

Learning and EHS awareness and competence should come from an environment of continual organizational learning emanating from top to bottom in an organization. Therefore the development of a *continual learning organization* is very important, which leverages on knowledge from its internal operations and lessons learnt as well as peer industries both locally and internationally.

There should be a clear set of expectations for safety and EHS at all levels, starting from the board. Also leadership explained that good EHS means good business; therefore, EHS should be, in the context of high risk, high reliability organizations, part of the business profile.

Transparency is critical to having a healthy working relationship between all stakeholders including the workforce, the management team, the leadership team and the board of directors. As discussed earlier, some leaders explained that transparency creates that safe and reliable culture and that it was leadership who created that culture of transparency by being fair and just. This is also consistent with a great deal of safety leadership and culture research (Odea and Flin 2001;[22] Conchie and Donald 2006;[23] Carrol 2002[24] etc.]. For example, the BP Texas Refinery Explosion Investigation (Baker et al. 2007, page xii)[25] in its report states: "A good PSM culture requires a positive, trusting and open environment with effective lines of communication between management and the workforce, including employee representatives."

Organizational resilience

Resilience is a very important aspect of high reliability. The level of preparedness of organizations to ensure both business continuity and emergency/crisis response planning was something that leaders said boards must expect of leaders of high risk organizations as a key expectation and objective. It was recognized in the interviews that focused efforts should be on prevention, but being prepared to respond to incidents and having action plans in place was equally important. It is important for the BOD to have clarity regarding the cost of resilience as an asset or as an expense when considering the balance sheet aspect.

On compliance matters the oil and gas leadership seemed to talk more about the *importance of internal controls and management systems*. Having good governance through internal management systems and processes should be the focus rather than on compliance to regulations, which is a given fact.

Ethics and morality

Some leaders touched on the fact that *ethics and morality of the leadership* is significant. The leaders interviewed from the construction industries said that this was a very significant aspect because they were competing with other contractors who may not have such high levels of morality and who may not as such treat their workers as well in order to simply cut costs and be more competitive in tenders. Whilst this may be a very straightforward point and there is considerable agreement on this, it is a fundamental aspect of creating trust between stakeholders, which is critical for a good EHS culture.

Independence of EHS

Finally, the **independence and strength** of the EHS function within organizations was discussed by some leaders, especially when it came to talking about reporting lines. The EHS function must be independent of other functions and report to the right level of command in an organization.

Regardless if the functional manager reported to the CEO/MD or not, the focus should be on the independence of the function in the sense that it could highlight EHS risks and issues relating to those risks without other parts of the organization directly influencing, interfering or changing their assessments of risks and impacts.

References

1 Al Hashmi, Waddah Ghanim (2011): "Safety Leadership Case Study – An In-depth Analysis of the Congressional Hearing With Tony Hayward, BP ex-CEO on the Deep-Water Horizon BP Oil Disaster, 2009 in the Gulf of Mexico, USA", Working Research Paper 01/11, Presented at 9th Annual OHS Congress, 5th–10th Feb 2011, Dubai, UAE.
2 Roy, Kaushik (February 2010): "SHE Management at West Kuwait Oil Fields of Kuwait Oil Company, Kuwait", ASSE-MEC-2010-48, American Society of Safety Engineers – Middle East Chapter Conference and Exhibition, Bahrain, Page 357.
3 Qadir, Tahir J. (2010): "Responsible Care – Integrating Security with QHSE", ASSE-MEC-2010-46, American Society of Safety Engineers – Middle East Conference and Exhibition, Bahrain, Pages 337–346.
4 Ahmed, Ismail M. (2008): "BAPCO's Experience in Implementing BBS Process", ASSE-MEC-0208-27, American Society of Safety Engineers – Middle East Chapter Conference and Exhibition, Bahrain.
5 Anderson, Greg (February 2008): "Creating a Culture of Safety", ASSE-MEC-0208-19, American Society of Safety Engineers – Middle East Chapter Conference and Exhibition, Bahrain, Page 131.
6 Roger, Isabela, Flin, Rhona and Mearns, Kevin. (2009): "Safety Leadership: A View of the Senior Managers' Role", Society of Petroleum Engineers (SPE), Paper presented in the 2009 SPE Offshore Europe Oil and Gas Conference & Exhibition held in Aberdeen, UK, 8th–11th September 2009.
7 Kakabadse, Nada K. and Kakabadse, Andrew P. (2007): "Chairman of the Board: Demographics Effects on Role Pursuit", *Journal of Management Development*, 26(2), 169–192.
8 Nicholson, Gavin J. and Kiel, Geoffrey C. (2004): "Breakthrough Board Performance: How to Harness Your Board's Intellectual Capital", *Corporate Governance*, 4(1), 5–23.
9 Deloitte (2008): "Perspectives on ERM and the Risk Intelligent Enterprise", The Enterprise Risk Management Benchmark Survey, Deloitte Development LLC.
10 O'Connor, Troy (February 2010): "SHE Leading and Lagging Indicator Dynamics", ASSE-MEC-2010-35, American Society of Safety Engineers – Middle East Chapter Conference and Exhibition, Bahrain, Page 273 (Ref 2).
11 Dunfort, Ghislain Giroux (2013): "The Risk Governance Imperative – Part 2", *Governance*, April (226), 5–9.
12 Van der Westhuyzen, Johan (February 2012): "Ensuring Contractor Alignment with Safety Culture", ASSE-MEC-2012-52, American Society of Safety Engineers – Middle East Chapter Conference and Exhibition, Bahrain, Page 372.

13 Gray, Garry C. (2009): "The Responsibilization Strategy of Health and Safety – Neo-Liberalism and the Reconfiguration of Individual Responsibility for Risk", *British Journal of Criminology*. May 2009, 326–342.

14 Organization for the Economic Cooperation and Development (OECD) (1st March 2012): "Corporate Governance for Process Safety – Draft Guidance for Senior Leaders in High Hazard Industries", Draft, Environment Directorate, Joint Meeting of the Chemicals Committee and the Working Party on Chemicals.

15 Health & Safety Executive (HSE) and Institute of Directors (IoD) (2007): "Leading Health & Safety at Work", Leadership Actions for Directors and Board Members, INDG417, HSE Books, Sudbury, Suffolk, UK.

16 Lofstede, Ragnar (November 2011): "Reclaiming Health and Safety for All: An Independent Review of Health and Safety Legislation", Presented to the Parliament by the Secretary of State for Work and Pensions by Command of Her Majesty, available from www.official-documents.gov.uk; Printed by the UK Stationary Office Ltd.

17 Clarke, Sharon (2012): "Safety Leadership: A Meta-Analytic Review of Transformational and Transactional Leadership Styles as Antecedents of Safety Behaviours", *Journal of Occupational and Organizational Psychology*. doi:10.1111/j.2044-8325.2012.02064.x.

18 MacAvoy, Paul W. and Millstein, Ira M. (2003): *The Recurrent Crisis in Corporate Governance*. Palgrave Macmillan, New York.

19 Al Hashmi, Waddah Ghanim (2013): "The Continuing Challenges of Balancing Process Safety Management and Personal Safety Management in the Oil and Gas Sector – The Role of Corporate Governance and Leadership", Caine Energy Global HSE Conference 2013, 26th–27th of September 2013, New Delhi, India.

20 Dermine, Jean (March 2011): "Bank Corporate Governance, Beyond the Global Banking Crisis", Working Paper, The INSEAD Working Paper collection, available from publications.fb@insead.edu, accessed 25th January 2014.

21 Jamali, Dima, Hallal, Mohammad and Abdallah, Hanin (2010): "Corporate Governance and Corporate Social Responsibility: Evidence From the Healthcare Sector", *Corporate Governance*, 10(5), 590–602.

22 O'Dea, Angela and Flin, Rhona. (2001): "Site Managers and Safety Leadership in the Offshore Oil and Gas Industry", *Safety Science* 37, 39–57, available from www.elsevier.com.

23 Conchie, Stacy M. and Donald, Ian J. (2006): "The Role of Distrust in Offshore Safety Performance – Risk Analysis", *Journal of the Society of Risk Analysis*, 26(5), 92–103.

24 Carroll, John S., "Leadership and safety in nuclear power and health care." Presented at the British Psychological Society Division of Occupational Psychology Scientific Seminar, University of Aberdeen, 21st May 2002.

25 Baker, James (III), Bowman, Frank L., Glen, Erwin, Gorton, Slade, Hundershot, Dennis, Levison, Nancy, Priest, Sharon, Rosentel, Tebo, Paul, Wiegmann, Douglas and Wilson, Ducan (January 2007): "The Report of the B.P. U.S. Refineries Independent Safety Review Panel", available from www.csb.gov.

11 Impact of board structures and future outlook on HSE governance

The final question asked to the senior leadership during the interviews was about the board structure's impact on EHS leadership and governance. As throughout this book, board structure and makeup as well as the relationship between the CEO/MD and the chairman and other board members were all highlighted as factors of success of the board. Therefore a hypothetical question was asked of these leaders on their opinion on which board structure they felt was best for EHS leadership and governance.

Figure 11.1 shows the distribution of views of the question of what kind of board was hypothetically best in terms of leadership and governance of EHS in organizations.

A total of five permutations were noted. There have been various debates around board structure and its impact on board performance,[1] and in this section we discuss the results with respect to this question in the context of EHS leadership. In response to this very question, it was interesting to note that the spectrum of opinions ranged greatly.

Key areas that should be considered when evaluating board effectiveness include the role, structure and composition of boards, as well as the processes and relationships of boards.[2]

Of the total respondents, 51.6% indicated a preference to a **mixed board structure** in which the CEO was a board member but not the board chairman or vice chairman. There were various reasons cited for this, which included that they believed such a structure brought about better alignment between executive management and the board members; it enables the CEO and perhaps some executives (who may also be appointed to the board) who are knowledgeable about the business to better contribute to the major decisions made which can impact on EHS and safety. They said that there is a better chance for the BoD members to understand the implications of the decisions they make, especially commercial and expenditure related, which can impact on EHS and safety; and that in general terms you end up getting better engagement from both the executive team and the BoD.

This is probably consistent with the views expressed by Bennington (2010)[2] and others such as Finkelstein and Mooney (2003),[3] who explain that whilst

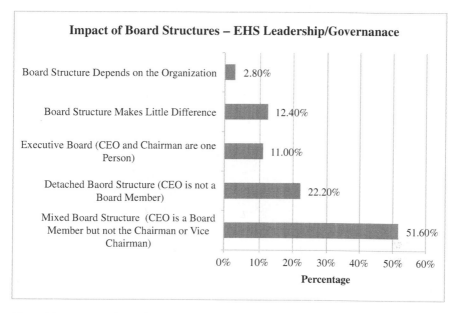

Figure 11.1 Impact of Board Structure of EHS Leadership and Governance

collaboration is required between the board and the CEO, the board's primary responsibility is to provide oversight, advice and monitoring of performance and, when needed, to counsel/discipline the CEO. More recent research, however, has indicated a changing landscape, with boards playing a greater role as strategic partners working in collaboration with the management team and now a changing, more collaborative leadership between boards and CEOs, especially in the health industry.[4]

There were also a relatively high proportion of respondents (22.2%) who supported the view of having a **detached board** in which the CEO and executive management only reported to the board and were not members. They cited many reasons, which included: (1) that there must be a clear delineation of the roles and responsibilities between the executive and the BoD and that the CEO is accountable for leading whereas the board were accountable for directing EHS matters; and (2) that this maintains, as they put it, better "Corporate Governance and Oversight", especially for safety issues, and especially when there is a potential in an organization to have major incidents. They also said that this ensured that the BoD makes more objective decisions because they are not influenced too closely by more subjective decisions from the CEO and his team, who may want to go ahead with a particular matter for a particular reason; and that this ensures that the BoD members remain as an independent directing and supervisory body.

This quote demonstrates this viewpoint:

> Yes, you must have that separation. I mean, governance is oversight, protection of the shareholders' interests and the company's long-range interests. Management is obviously focused on the next quarter, on the next year, they're results driven. I think separation is good.
>
> (Respondent 22, Oil and Gas, Upstream, page 22[5])

Only around 11% of the respondents supported having an *executive chairman board* in which the CEO and chairman were one. They cited their reasons for supporting such a board structure by saying: (1) If EHS starts at the top, then the chairman, as the CEO, will ensure EHS issues are managed with no compromise; (2) they also said that there would be no issues with misalignment between the BoD and the executive team, as they will be one and the same, and that the BoD would become very effective advisors for ensuring the decisions that are made are balanced when it comes to EHS issues. "Duality" of roles of a CEO also as a chairman has become quite a controversial issue, although it is common (and thus an acceptable practice) in countries like the USA, Hong Kong, North Africa and the Arab world. Furthermore, in a review of the corporate and healthcare governance literature Lynne Bennington explains: "Whilst agency theory views duality as inappropriate, as it reduces the monitoring role of the Board, stewardship theory views duality as removing any ambiguity about who is directing the organization and, thus, sees it positively" (Bennington 2010, page 321).[2]

To this end, others have also indicated that executive duality should not exist and that a clear separation should pertain of the head of management and head of the board, and this to the extent that even an executive vice chairman position would not be acceptable in the context of the banking industry to ensure effective governance and oversight.[6]

Some supported an either/or approach – i.e. both *executive and mixed* and *detached and mixed models*. However, interestingly, 12.4% were of the view that board structure made very little difference. This was for a few reasons cited such as:

1   That safety and EHS is about responsibility and accountability, so it is not the board structure that will be the reason for better or worse safety leadership and culture within an organization;
2   The performance of the board in terms of EHS matters depends more on the optimum number of board members who have varied experiences rather than structure; and
3   That this depended more on the general diversity of the board rather than the structure of the board, and that EHS was ultimately all about people.

They also said that the board structure will continue to depend more on regulations and international best practice, and safety/EHS matters would be absorbed within those structures.

Board structures are very much influenced by size and type of the organization, as clearly expressed by a CEO of a manufacturing company who is also a corporate governance specialist; he explains: "Now when you speak about board structure you have to think about them. It is a structure that has to be in compliance with the regulations and law for setting up the company" (Respondent 18, Manufacturing, page 13).[7] And also another aspect from a shareholder point of view was that they set up the board according to the size and type of function that the organization is doing; the following quote illustrates this: "So if you have a small organization which consists of four guys, you don't have to go through the trouble of setting a whole full-fledged board" (Respondent 18, Manufacturing, page 13). This is to some extent also consistent with the views of Millstein and MacAvoy (2003)[8] and Bauer and Schmidt (2008)[9] etc.

Finally, one respondent did express that board structure effectiveness depended more on the organization and the relationship dynamics between the CEO and the BoD.[10]

References

1 Verdeyen, Vanessa and Van Buggenhout, Bea (2003): "Social Governance: Corporate Governance in Institutions of Social Security, Welfare and Healthcare", *International Social Security Review*, 56(2).
2 Bennington, Lynne (May 2010): "Review of the Corporate and Healthcare Governance Literature", *Journal of Management and Organization*, 16(2), 314–333.
3 Finkelstein, Sydney and Mooney, Anne C. (2003): "Not the Usual Suspects: How to Use the Board Process to Make Boards Better", *Academy of Management Executive*, 17(2), 101–113.
4 Bjork, David A. (2006): "Collaborative Leadership: A New Model for Developing Truly Effective Relationships Between CEO's and Trustees", Centre of Healthcare Governance.
5 Interview with a CEO of a GCC-based oil and gas Upstream Company – Conducted in May 2013 in Sultanate of Oman.
6 Olayiwola, Wumi K. (November 2010): "Practice and Standard of Corporate Governance in the Nigerian Banking Industry", *Journal of Economics and Finance*, 2(4), 178–189.
7 Interview with a CEO of a GCC based Manufacturing Company – Conducted in April 2013 in Abu Dhabi, United Arab Emirates.
8 MacAvoy, Paul W. and Millstein, Ira M. (2003): *The Recurrent Crisis in Corporate Governance*. Palgrave Macmillan, New York.
9 Brauer, Matthias and Schmidt, Sascha L. (2008): "Defining the Strategic Role of Boards and Measuring Boards' Effectiveness in Strategy Implementation", *Corporate Governance*, 8(5), 649–660.
10 Interview with a CEO of a GCC based Aviation Sector Organisation – Conducted in March 2013 in Dubai, United Arab Emirates.

12 Model of EHS leadership in the context of corporate governance

The areas or themes representing the aspects of this leadership and governance with respect to EHS/safety are represented within a contextual diagram. With the themes emerging from both the academic and practitioner research and interviews with senior leaders, the diagram just represents where these areas exist. Thus the conceptual model is populated with the themes, transforming it into the **Contextual Diagram of EHS Leadership and Governance**. The diagram is represented in Figure 12.1.

The model shows the three areas of the research. The challenge of this research has been to find the areas where these three spheres of knowledge and practice overlap. The model shows equal triangles, but this is misleadingly simplistic, as the reality is much more complex, and thus in practice the triangles may be different in their areas and degree of overlap. However, what we know well is that the areas of overlap between EHS and leadership can be defined by the EHS/safety leadership theories and models which have been explored mainly by EHS practitioners and perhaps industrial psychologists. The area of overlap between EHS and governance is an area defined as compliance, whereas the area between governance and leadership can be defined as the performance standards which define how the board evaluates CEO/MD performance and initially sets expectations.

The themes fall within different areas within this conceptual diagram. The themes, as already explained here particularly and in the earlier chapters generally, are distinct yet very much interconnected. They influence each other and drive one another's importance and emphasis, yet they can still be explored as exclusive themes as well. Because the subject is relatively complex, with so many factors, and to better understand it holistically, an understanding of the themes and their positioning within the contextual diagram helps define the model, which is presented further for that purpose and utility.

Because of the complexity of this kind of study due to the many themes (or factors) to be conceptually understood as a whole, by understanding where the themes lie in the diagram, we better understand the subject as a whole.

The themes which are given in light grey are the themes that initially emerged from the literature review. We find that only one theme seems to sit comfortably within the very centre – notably the monitoring of EHS performance. Another

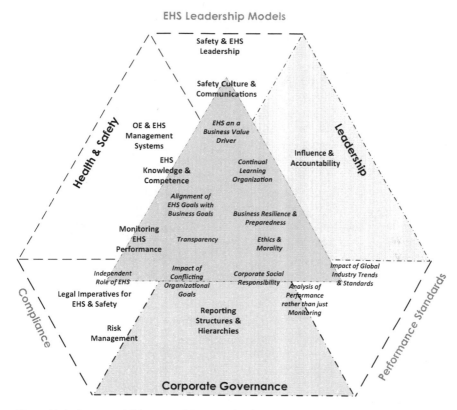

Figure 12.1 Contextual Diagram of EHS Leadership and Governance

three themes sit near the borders; safety culture and communications fall within the border between the centre and EHS leadership, distinct yet very much connected to EHS leadership, and, as discussed earlier in this chapter, in fact culture can be said to be a product of that leadership.

Knowledge and competence and **influence and accountability** also border the central overlap area with performance standards. Boards expect a certain level of influence on organizations' workings and also expect that accountability is taken for the results, whether positive or otherwise. The same goes for EHS knowledge and competence, only here this is applicable as influence and accountability to both the executive and the board leadership.

The other themes explored also fall within the other areas such as compliance, risk management and the legal imperatives for safety. OE and EHS management systems also are within the area of compliance, but due to the technical nature we see the overlap with EHS. This explains the significant variation between the respondents that we saw earlier in the results with risk management and OE and EHS management systems, and also between the oil and gas and non-oil and gas industries. Whilst these fall within the compliance theme, they are not within the zone of "performance standards setting" (i.e. the area between governance and leadership), so due to the current board's nature and EHS knowledge and competence, they see these two themes being addressed by influence and accountability.

With respect to **EHS reporting structures and hierarchies**, there were differences in responses, with reporting lines of senior EHS managers/directors within organizations sometimes reporting directly when senior leaders saw: (1) that they needed to be hands-on with EHS issues; (2) that it was important to promote the role of EHS in a message manifested by the direct reporting line to the top person within the organization; or (3) simply that the reporting of EHS had to be direct to ensure speed and quality of reporting of information relating to EHS as a critical organizational matter. This is consistent with McLean (2003), who stressed that practitioners must report to the organizational leaders so that they may help in directly implementing their commitments but also as this reporting line in itself demonstrated to everyone within the organization that this position is a senior one, carrying equal importance to all the other operational, technical and financial functions.

On the other hand, some other senior leaders interviewed saw that the reporting lines made little difference as long as there was a structure, strong procedures and management systems with responsible senior technical directors reporting to them, where EHS matters were raised through the operational and technical management layer. Overall, this theme is a matter of corporate governance.

On reporting structures, some of the leaders interviewed felt that as long as there is an independent role for EHS within the organization, the reporting was a secondary consequence. This emergent theme, it is felt, was within the border of compliance and governance.

As discussed in the previous section, the other emergent themes (shown in italics) are also linked to one another and to the themes that existed at the start of the research. However, they have been developed through the iterative process undertaken to bring together common aspects of the string of ideas that emerged from reviewing the interviews undertaken. There is no right or wrong theme, and they could have been combined in other ways. The researcher here used experience and reflective knowledge over 20 years of experience in the field of both EHS and management to bring these different ideas together to create these newly emerged themes.

In saying this, they may not be themes but important matters to understand in this developing model. For example, much of the ideas that emerged from the interviews came from the reflections of senior leaders, and, notwithstanding the fact that much of the literature was developed by practitioners in various fields of knowledge such as legal, leadership, safety, EHS, governance and so on, these themes or ideas are real. They are what senior leaders see and feel and have an impact on this whole notion of EHS leadership in the context of governance.

With performance standards, the impact of conflicting organizational goals, the impact of the standards within industry and the ethics, morality and corporate social responsibility were also aspects driven by the relationship and expectations set by the board and the CEO/MD. The interviews very much revealed, especially when leaders were asked to comment on their future outlook on governance and EHS leadership, both internal and external drivers set an agenda and thus set the performance expectations from the executive management. A great deal of impact, it was discussed, has been produced by the rise of industry standards in the past few decades. This is particularly true with the oil and gas industry but more pronounced in the aviation sector.

On corporate social responsibility, at least three highly influential and senior leaders explained that the very viability and sustainability of their businesses, in which the shareholders were governments, was not possible without their organizations continuing to contribute to creating jobs for the nation's workforce; socio-economic development through sponsorships; developing local talent; awarding contracts to local businesses and generally supporting society at large. This, the researcher felt, needed to be translated into performance expectations by the board or the CEO/MD. Whilst it is related to EHS, especially when it comes to labour rights and environmental stewardship, it is sufficiently distanced in the model to sit where it does under performance standards.

Then, at the very heart of this contextual diagram sit many important ideas or themes that arose from the senior leaders' qualitative inputs. Some of these themes are organizationally tactical, and others the researcher views as highly strategic. The more organizational themes are those which related to ***transparency*** as a culture where there is open reporting and generally open communication systems without the fear of reprimand, which is also linked to leadership and the creation of a just and fair culture, as discussed earlier in regards to[1] high reliability organizations that have a ***just culture***, promoting transparency in reporting of incidents and improvements with a great balance between supporting the reporting culture and tolerating unacceptable behaviours.

Also at the heart of the diagram, the ***ethics and morality*** of leadership play a very important role in EHS leadership and governance. EHS is mainly about the protection of people, environment and assets and loss prevention. This does not always amount to pure financial rewards. In fact, sometimes operating in a risky way even if there is a high potential loss may mean very high returns, and without basic ethics and morality it may otherwise be justified to operate in an unsafe way as long as you do not get caught.

Linked to EHS knowledge and competency at an organizational level is having a ***learning organization***. This is also linked to safety culture and communications as well as transparency. Many interviews demonstrated that EHS and improvements that are being made in organizations were part of a long learning journey that the whole organization including the board was on. In that journey, continually learning was pivotal to improvement and the eventual running of operational excellent EHS management systems.

Analysis of performance may be combined with the ***monitoring of EHS performance*** theme, and in the final model we see this as an outcome of the many themes and factors that have been extensively discussed. The themes were kept separate for the purpose of illustration; various senior leaders interviewed explained that boards should analyze performance and not just monitor. Their main arguments were that if the information presented was analyzed, they would be actually playing a more engaged role, effectively understanding what is going wrong but also what is going right and thus better appreciating the efforts being exerted by the whole organization to maintain a high level of performance.

On *alignment of EHS goals with business goals*, this theme relates strongly to a consequential theme being developed which also resides in the heart of the diagram, which is making EHS an *organizational value driver*. It was recognized by some of the senior leaders that this misalignment between EHS and business goals created the lack of *investment in EHS* and also, as discussed earlier, created the *impact of conflicting organizational goals*. So in their own right these aspects or themes have a strong bearing on the model, and thus perhaps when looking at solutions to improve governance and EHS leadership in organizations, these aspects must be studied carefully, and pragmatic solutions need to be presented. By making EHS a core business/organization value driver, its importance is presented to everyone within the organization, including the board. Only here the board must drive this through setting an expectation of the organization to make EHS a core value. This is what the best-practice guideline[2] explains about the BoD setting the tone for the organization, and in doing so, they were demonstrating their EHS leadership.

Finally, there has been much discussion about high reliability organizations and their attributes and also sustainability in earlier chapters. During the interviews the overarching paradigm that was felt about EHS from the business leaders was based on having policies and systems in place to prevent incidents from occurring and being ready to deal with incidents when they do occur. Thus, *business resilience and preparedness* sits also in the heart of the diagram. In the context of high risk industries, to become high reliability organizations, business resilience and preparedness becomes a theme of both significant and central importance.

References

1 Health and Safety Laboratory (HSL) (2011): "High Reliability Organizations – A Review of the Literature", Health and Safety Executive Research Report – RR899, HSE Books, Sudbury, Suffolk, UK.
2 HSE/IoD/LGE (2009): "HSE/IoD Guidance – A Report on the Awareness and Use of the Leading Health and Safety at Work Publication Within the Local Authorities", Local Government.

13 The model of EHS governance and leadership

In Chapter 12, the Contextual Model has helped in synthesizing the themes and thus brings together all the factors in the form of themes which have come from both the reviews and the interviews undertaken.

However, the diagram is too fluid and fails to give us three fundamental requirements:

1   The first is the relationship of the themes and factors with one another, as they surely are linked and influence each other;
2   Second, from a very basic analysis the diagram is not able to arrange these factors into groups; and
3   Finally, the conceptual diagram, although it places factors within the research areas explored, fails very much to link these factors in such a way as to explain their ultimate impact on organizational EHS leadership and governance.

So whilst we should be confident that the diagram provokes further thought, creates an excellent foundation for debate and, most importantly, in an abstract way positions this highly complex set of inter-related themes and concepts, subsequent to its creation a more meaningful and practical model has been developed.

Figure 13.1 builds on the model given in Figure 12.1 in Chapter 12 and defines the themes that emerged in this review work.

The themes have been grouped in to four key areas, including internal organizational factors; external social, political and economic factors; personal leadership factors; and finally enterprise business factors.

As we have already explained the factors, the reader is invited to note the key developments as stated hereunder to understand these themes:

Internal organizational factors

Leaders, when probed on the three areas that included operational excellence and management systems; reporting structures and hierarchies; and safety culture and communication, confirmed their importance in their responses.

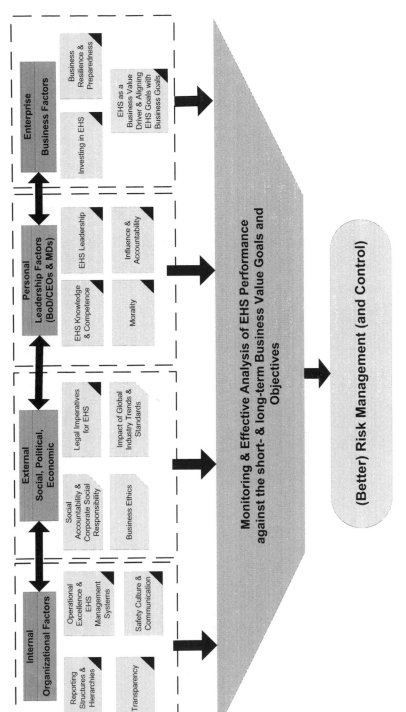

Figure 13.1 Model of EHS Leadership and Governance

However, there is an additional factor or theme on transparency which also evolved from the inferences from the discussions undertaken. Transparency has much impact on improving the trust within an organization, and therefore leaders must promote transparency starting with themselves. The previously listed four factors have been classified or grouped under the internal organizational factors.

External social, political and economic factors

Perhaps what was not fully appreciated when external factors were initially studied at first glance was that we sometimes fail to highlight the factors that relate to business ethics; social responsibility and accountability and, most significantly, the influence of global (trends) practices and standards have a significant impact within the external factors. The legal imperatives for safety that much of the literature emphasized were important and in fact seem to always take utmost importance.

This is not the case today; now aspects relating to the pressure of industry for organizations to maintain and improve their operational EHS standards are significant. What is important to consider in this model is that factors that relate more to leaders demonstrating their organizations as being ethical, good social enterprises and good corporate citizens is very important. These four areas may be the most important external factors on a business which wants to demonstrate effective EHS leadership and governance.

Personal leadership factors

The aspects that relate to good EHS knowledge and competence, EHS leadership and a self-realization of a leader, be they at the CEO/MD or board director level of their influence and accountability, have been greatly discussed in earlier chapters. In fact the influence that leaders have over the organization is very important indeed. Communicating effectively on the values and on the proposition that EHS and good practices have a positive effect on an organization's long-term sustainability is such a critical leadership aspect.

The morality of leadership also came out as a critical personal factor from many of the interviews. Morality is an internal factor which drives a leader's behaviour, and whilst some (limited) literature mentioned the aspects of morality, there is not enough in the current literature which truly highlighted its importance, as did the interviews with senior decision makers.

Enterprise business factors

One of the most important aspects of the leadership model – and possibly as a consequence of the disciplines of EHS leadership and governance that are currently researched and are written about in this space, which the literary review was based on – is the impact of the enterprise factors.

These business factors, which were illustrated in the interviews, were in three key areas. EHS is seen by the leaders of high reliability organizations as a business value driver, and therefore EHS is simply good business in their minds, and it very much is because it drives costs down, reduces losses and is directly positively correlated to good productivity. Moreover, EHS matters need to be effectively aligned with business goals and objectives, and this is an important influencing factor, especially when one appreciates that, at the very basic level, EHS provides a long-term sustained business imperative.

This in turn also leads to the fact that business continuity through reliance and preparedness for incidents is an important enterprise aspect to consider. Lastly, and a very important and refreshing viewpoint that many leaders expressed, was that with EHS they saw an investment, and thus it was important to invest in safety for the long-term healthy growth of the enterprise. This is opposed to the more published view of EHS being purely a cost of compliance.

Oversight = monitoring, analyzing and managing risk effectively

When we go back to the basic responsibilities of a board director, monitoring company performance is a key aspect. With this viewpoint, directors can help redirect, re-evaluate and address any matters that need a strategic relook to ensure alignment or improvement of company performance.

The key issues explained in many references and in fact confirmed very much in many interviews with senior leaders is that whilst board directors took a great interest in company performance metrics and monitored them against targets, they were less involved in asking critical questions which analyzed the reasons for that performance, be it good or bad. EHS performance monitoring at board level is not in any way different. In fact, less analysis is probably undertaken by the board, given that EHS matters are seen by many as very technical.

So it is not simply the monitoring of EHS performance as much as it is the monitoring and effective analysis of performance that was expected by the board that adds value to the mid- and long-term business value goals and objectives. Thus, board directors must get more involved and make a greater effort to understand reasons for EHS performance to guide and direct the executive team so that they can manage it effectively.

The model describes that all these themes that relate to leadership – internal organizational, external business environmental and enterprise factors – impact directly on risk management and can thus be predicators of effective risk controls that can in turn effectively ensure the long-term sustainable growth of high risk/high reliability organizations.

What the model also clearly demonstrates is that effective control can be achieved through effective monitoring, meaningful insights born from a better holistic understanding of all the themes that relate to EHS leadership and governance leading to a more effective governance of EHS.

The model also highlights the actual complexity in terms of the number of factors and sub-factors that influence the effectiveness of monitoring, analyzing and managing risks effectively. These factors need to be considered in leadership development programmes and when we look at influencing the development of effective risk management in board practice.

14 Conclusions

The demand for energy, construction, travel, logistics and industry in general is increasing. There is a growing population, and many of the large high risk/high reliability organizations are either government-owned or otherwise major privatized infrastructure entities that are strongly associated with national interests. They are huge employers directly and also significant influencers in the development of other supporting service industries which also play a vital role in creating jobs, social value and long-term income and economic prosperity.

The importance generally of corporate governance globally has increased in the past 15–20 years, and in more recent years corporate governance standards have recommended going beyond financial reporting to reporting the non-financial performance of organizations. With a fast-changing world and greater complexities, the function and effectiveness of the board of directors as a body has come under greater scrutiny.

Organizational leaders' role in defining and continually improving the safety and EHS culture is vital. Just, fair and transparent organizational cultures play a very big role in ensuring better EHS cultures, and EHS in general should be led from the top.

The development of EHS management systems, operational excellence systems and overall the integration between organizational EHS management systems emphasizes the importance of the management system as an effective tool to maintain and improve EHS performance. Legal and regulatory imperatives are increasing. One of the key roles of internal policies/standards of practice in an organization is that they help ensure compliance with statutory regulations. Whilst it may be debatable to what extent MDs/CEOs and board members are responsible for omissions of shop-floor employees, executive and board management must ensure appropriate policies, processes and procedures are in place to reduce risk and ensure compliance with both regulations and good practices from the industry.

The impact of leadership on the performance of organizations and their sustainable growth is critical. Sustainability has become a central theme in organizations. Asbury and Ball (2009)[1] give an excellent overview of the development of this aspect of sustainability, focusing on the rise of corporate social responsibility (CSR), which was already discussed in both Chapters 3 and 7 as an important component of organizational dynamics.

Among the negative impacts related to EHS they cite are habitat destruction, use of resources, waste generation, noise, local safety issues and other pollution issues. They explain how stakeholder expectations have started to set a tone for overt organizational behaviour and where stakeholders other than shareholders have begun to have a greater impact. They identify five main types of stakeholder including (1) customers; (2) employees; (3) suppliers and contractors; (4) shareholders; and (5) society at large. Of course, within society non-governmental organizations (NGOs) have in more recent years played a very important role, especially after certain environmental and safety incidents where significant damage like environmental pollution and/or fatalities/injuries has occurred.

Many states within the Middle East and around the world, especially in the emerging markets, require a good and solid framework going forward for good EHS practice and governance. This is particularly important with fast-changing socio-economic drivers where the regional economies rely heavily on high risk/high reliability businesses. This book contributes to advising of the foundations of those frameworks that must be developed.

As an example, the BDI-GCC,[2] as discussed in the earlier chapters, was established to develop and improve and influence the standard of boards in the GCC. Whilst they have not really looked at EHS issues to date, they have in their programmes addressed both financial and non-financial performance review and management. Their programmes also focus on highlighting the legal and fiduciary duties that boards carry and some of the aspects on the legal imperatives for EHS.

The developments in corporate governance have become more profound with the economic challenges that have been faced in many parts of the world. This serious sense of ineffective corporate governance and control, specifically in the context of financial issues, has driven a transition from management system control philosophy in EHS to more HRO development, stewardship and engaged leadership.

Much of the recent work on HROs has highlighted the importance of a more engaged leadership role in organizations. Many of the new corporate governance standards, especially those highlighted by the IoD, UK-HSE and OECD, have started to talk seriously about EHS risk management in a more pronounced and overt fashion, emphasizing the need for leadership, both at the executive and board level, to play that engaged role. Whilst much of this development has been from a Western context, these frameworks with respect to the legal context influenced many internationally based corporations working in the GCC area. These corporations have a direct influence on boards, especially in the joint venture (JV) and joint stock companies, including many energy and energy-related companies such as the Abu Dhabi National Oil Company (ADNOC), which has various shareholdings with oil majors in their different production subsidiaries, Tatweer (a JV between the government of the Kingdom of Bahrain,

Oxy and Mubadallah from the UAE), ASRY (a major heavy construction JV involving five different Arab states) and many similar examples of large high risk operation companies in the GCC region.

In that sense the EHS leadership research and that relating to safety culture development in organizations continue to really drive this point very strongly, going to the extent of saying there cannot be really effective control and development in an organization without serious leadership involvement and engagement. Boards have to get involved in monitoring and managing EHS risks. To this end, leading from the top and achieving a company-wide buy-in and understanding process safety management (PSM) will decrease future financial liabilities (Fowler 2011).[3]

The role of the board chairman as a chief risk governor and the CEO/MD playing the role of a chief risk officer has become more pronounced with corporate governance codes of practice. The relationship between these positions is important, and the board members' competence and board structure all play a role. There is very little in the way of reviews and published papers about corporate governance and EHS leadership/EHS management systems. There are a few anchor references which have been discussed in this book such as the joint HSE-IoD and the OECD publications. Their review is very much part of the fiduciary duties, regardless where the enterprise actually physically operates.

In saying this, there is evidence that in more progressive companies (as seen in more recent best-practice research), organizations seek directors with relevant industry experience with expertise when recruiting new board members (PWC 2010).[4] This includes those who have a better understanding of the global environmental challenges, social responsibility and sustainability matters.

The legal and regulatory imperatives and the more recent developments in taking persons in senior organizational roles and even boards to task are extremely insightful and will help shape a more informed leadership in the future. What must be explored further is to what extent these developments have brought about changes in the corporate leadership's approach to wanting to understand more. In the GCC there are also interesting and rapid developments towards greater regulation and establishing of accountability. As discussed in the chapter covering laws and regulations in the more emerging markets such as the GCC, EHS laws and regulations – whilst becoming more structured and prescriptive to address EHS, particularly from the perspective of the requirements for organizations to establish management systems – also very much highlight more and more the appreciation of effective risk assessment and management at organizational decision-making levels.

With increasing numbers of global cases being tried, the degree of foreseeability in the industry will be put to the test in the future. Organizations will have to argue that the best practices they instituted are in fact so and that they have learnt from other global incidents. Furthermore, as the lawmakers and jurists are learning more about the semi-technical concepts of risk management, boards and executives failing to demonstrate that they have been diligent in preventing losses will subsequently become more prone to potential liabilities, both as organizations and as individuals.

From becoming more green to demonstrating their responsibility as socially responsible organizations towards their employees, contractors and society at large, organizations are having to play a proactive role in presenting themselves in that positive light.[5] Having EHS incidents where lives are lost, significant damage to the environment is caused or other major problems occur is very damaging to the bottom line and erodes shareholder value. Moreover, the expectation from leadership to raise the bar and expectations to demonstrate what they did to prevent something like this from happening is now for them a fact of life.

Throughout this book, and especially in the last few chapters, the author has also stressed effective communication, being informed and the importance of effective reporting lines. The role of the EHS practitioner has changed from the discipline engineer or practitioner to an important technical risks advisor to the leadership, be they executive managers or directors. Their role in prevention cannot be overstated, given their knowledge and specialist expertise.

It is clear from the research evidence that the concepts in ERM have become the new way that boards and executive leadership are able to manage risks in what have been established as some of the more serious high risk/high reliability industries.

Ultimately, risks are an essential part of doing business, and setting the appetite for risk and risk control strategies will help ensure organizations meet their objectives.[6] To add value, however, companies must go beyond compliance and look at how risks can be integrated into every significant decision – in other words, the creation and dissemination of a sound risk culture – and thus be aware of risk and leverage it correctly. No doubt that the board director must be a strategist as well as a constructive challenger and be aware of his/her fiduciary and higher organizational interests. Board directors should be "competent" enough to understand the kind of EHS risks to add value to an overviewing of an organization's performance over time in order to deliver a sustained business operation.

Serious and systemic consideration of risks was of equal importance to the drive towards increasing performance and in general good governance. The role of the audit committee should be separated from that of the role of the risk committee, which should have a different approach to looking more at enterprise risk management and focus outwards (externally) on emerging and dynamic risks rather than the internal risks of non-compliance.[7]

Ultimately, "managing a major hazard business should be a clear and positive process safety leadership with board level involvement and competence to ensure that major hazard risks are being properly managed" (HSE 2011, page 11).[8]

It is critical that we therefore explore all these aspects with actual CEO/MD and other top leadership in organizations directly to understand how much their views truly are aligned with the many academics, leadership development specialists, governance experts and EHS practitioners' views presented in this book.

Final thoughts

Serious and significant challenges face the senior leadership in operating high risk organizations of the future. The direction, development and long-term sustainability of major organizations which serve major interests such as the energy, transportation, construction, manufacturing and other such industries depend today more than ever before on effective leadership actions. The enterprise risk oversight, the effective risk governance and the alignment of organizational business goals with EHS, social accountability, sustainability and other such goals has become a very serious consideration that every senior executive and board director should appreciate.

The cost of getting it wrong in a high risk industry has the detrimental effects of eroding company value in potentially such a short period of time, and this, compounded by the legal liabilities for ineffective risk control and governance, must make senior leadership at both an executive or non-executive capacity think very carefully when making decisions, not to mention the potential macro-economic impacts at a regional and global scale.

This book is a small yet a very comprehensive contribution to the world of effective risk governance and EHS leadership in modern and future high risk organizations aspiring to become high reliability organizations.

References

1 Asbury, Stephen and Ball, Richard (2009): *Do the Right Thing – The Practical, Jargon-Free Guide to Corporate Social Responsibility*. Institute of Occupational Safety and Health (IOSH), Leicestershire, UK.
2 Gulf Cooperation Council – Board Directors Institute website, www.gccbdi.org.
3 Fowler, Andrew (September 2011): "Why Taking an Active Role in Safety Is Good for Business", *Tank Storage Magazine*, Regulations Section, 45–46.
4 PricewaterhouseCoopers (PWC) (May 2010): "Boards Respond to Stakeholders With Stronger Oversight", Corporate Governance Trends and Best Practices Series – Global Best Practices, PricewaterhouseCoopers Publications.
5 Phyper, John-David and MacLean, Paul (2009): *Good to Green – Managing Business Risks and Opportunities in the Age of Environmental Awareness*. John Wiley & Sons, Canada.
6 Dunfort, Ghislain Giroux (2013): "The Risk Governance Imperative – Part 2", *Governance*, April(226).
7 Gregory, Stephen (2011): "Risk Management and Insurance: From the Guardroom to the Boardroom", Financier Worldwide – Reprinted From Global Reference Guide – Risk Management and Insurance.
8 Health & Safety Executive (2001): *A Guide to Measuring Health and Safety Performance*. HSE Books, Sudbury, Suffolk, UK.

Index

Note: Page numbers in *italic* indicate a figure and page numbers in **bold** indicate a table on the corresponding page.